GREENER PASTURES:

Decentralizing the Regulation of Agricultural Pollution

As farms grow in size and become increasingly industrialized, the problem of agricultural pollution is becoming more and more urgent across Canada. The response from most environmentalists and provincial governments has been to push for more centralized regulation. In *Greener Pastures*, Elizabeth Brubaker exposes the detrimental effects of such centralization, which has tended to exacerbate rather than curb pollution.

For centuries, Brubaker explains, conflicts about farming were resolved by the parties directly involved, aided by common-law courts. The rule, 'use your own property so as not to harm another's,' fairly and effectively resolved disputes between farmers and their neighbours and curbed environmental damage. Beginning in the 1970s, however, concerns about restraints on agriculture's growth prompted governments to replace the common law with more permissive provincial statutes.

Greener Pastures chronicles the centralization of agricultural regulation and the resulting environmental harm. Brubaker focuses specifically on the 'right-to-farm' laws (passed by every province in recent decades) that have freed farmers from common-law liability for the nuisances they create. She shows how these laws have made possible an unsustainable intensification of agriculture, and argues for a decentralized, rights-based decision-making regime. This thoroughly researched and impressively developed study challenges many common assumptions about environmental regulation, and proposes fresh answers to urgent environmental and political questions.

(University of Toronto Centre for Public Management Monograph Series on Public Policy and Public Administration)

ELIZABETH BRUBAKER is the executive director of Environment Probe.

The University of Toronto Centre for Public Management Monograph Series

Editor: Andrew Stark, University of Toronto

The University of Toronto Centre for Public Management Monograph Series is an ongoing series of books on important topical matters in public administration and public policy that will engage not only the academic community, but also policy- and opinion-makers in Canada and elsewhere.

Books are included in the series based on their originality, capacity to provoke public debate, and academic rigour.

For a list of books published in the series, see p. 155.

ELIZABETH BRUBAKER

Greener Pastures

Decentralizing the Regulation of Agricultural Pollution

University of Toronto Centre for Public Management Monograph Series

© University of Toronto Press Incorporated 2007
Toronto Buffalo London
Printed in Canada

ISBN 978-0-7727-8621-0 (cloth)
ISBN 978-0-7727-8620-3 (paper)

Printed on acid-free paper

Library and Archives Canada Cataloguing in Publication

Brubaker, Elizabeth, 1958–
    Greener pastures : decentralizing the regulation of agricultural
pollution / Elizabeth Brubaker.

    (University of Toronto Centre for Public Management monograph series)
Includes bibliographical references and index.
ISBN 978-0-7727-8621-0 (bound)
ISBN 978-0-7727-8620-3 (pbk.)

    1. Agricultural laws and legislation – Canada.   2. Agricultural
pollution – Canada.   3. Agriculture and state – Canada.   I. Title.   II. Series.

KE1679.B78 2007   343.71'076   C2006-906993-X   KF1682.B78 2007

Printed for the University of Toronto Centre for Public Management by the
University of Toronto Press.

Contents

Foreword

Over the past few decades, Canadian farms have increased in size and intensity. As a result, agricultural pollution – especially the contamination of groundwater and surface water and the emission of noxious gases and other airborne contaminants – is now a contentious legal and political issue all across Canada. Environmentalists and provincial governments have typically responded by pushing for more centralized regulation. In *Greener Pastures*, Elizabeth Brubaker argues that such regulatory changes bring perverse results. They tend to exacerbate rather than curb pollution.

For centuries, Brubaker explains, conflicts about farming were resolved by the parties directly involved, aided by common-law courts. The rule, 'use your own property so as not to harm another's,' provided the foundation for resolving disputes, balancing the interests of farmers with those of their neighbours. This regime – guided by firm principles and precedents, yet site-specific and adaptable – fairly and effectively controlled agriculture's adverse impacts.

Beginning in the 1970s, governments, wanting to promote agriculture's growth, began replacing the common law with more permissive provincial statutes. Brubaker chronicles the centralization of regulation and the resulting environmental harm. She focuses on the right-to-farm laws (passed by every province in recent decades) that have freed farmers from common-law liability for nuisances. Such laws often moved disputes about agriculture's impacts to administrative bodies charged with assessing not whether farming practices harm others, but whether they are 'normal.'

Establishing normalcy as the test of acceptability, Brubaker demonstrates, has made possible an unsustainable intensification of agriculture. It has created an industry whose costs are borne by those living downwind and downstream. Brubaker proposes a return to a decision-making regime that internalizes costs, creates incentives for farmers to operate sustainably, and respects the principle of subsidiarity, which places decision making as close as possible to affected citizens.

Greener Pastures proposes fresh answers to such questions as Who should decide what amount of pollution is acceptable? Who has the strongest incentives to ensure that practices are sustainable? What institutions best protect the environment? When are local solutions desirable, and when are provincial or federal solutions called for? A timely entrant into the public debate, *Greener Pastures* challenges many common assumptions about environmental regulations, assumptions long held by the environmental community, agricultural groups, and provincial governments.

Andrew Stark
Centre for Public Management
University of Toronto

Acknowledgments

This book owes much to the research assistance provided by colleagues and volunteers over the last several years. I thank Robin Longe for his research into the common law; Maya Paul, Kate Tsiplova, and Lindsay Forbes for their examination of right-to-farm legislation and the decisions of right-to-farm boards; Kate Tsiplova (again) for her work on municipal planning; Deborah Duffy, Ailish Murphy, and Elliot Siemiatycki for their research into federal regulatory mechanisms; Lisa Robles for her work on the environmental impacts of farming; Tobey Ann Pinder for her exploration of the positions taken by environmental groups; Ted Cooper and Tom Adams for their insights into drainage; Geoffrey Patridge for his toil at the library; and Richard Owens for his unfailing willingness to track down elusive cases and discuss legal intricacies.

I am also indebted to Larry Solomon for his comments on both a preliminary draft of the book and the completed manuscript, and to editor Andrew Stark and two anonymous reviewers for their comments on the manuscript. Their economic and legal insights were extremely helpful.

This book was made possible by donations from Environment Probe's individual supporters and by grants from the Earhart Foundation (administered by the Property and Environment Research Center), the Helen McCrea Peacock Foundation (administered by the Toronto Community Foundation), and the Margaret Laurence Fund. I am grateful to all for their generous support.

GREENER PASTURES

1 Canada's Farmers: Salt of the Earth or Assaulting the Earth?

Current farming practices are not sustainable.
 – Canada's Environment Commissioner, Johanne Gélinas, 2001[1]

In May 2000, a deadly strain of E. coli bacteria contaminated the drinking water in Walkerton, Ontario, killing seven people and sickening more than 2,300. Scientists traced the bacteria to a small cattle farm in the community. They deduced that a heavy rainfall had washed the bacteria from manure spread on the farmer's fields and sent it, through cracks in the soil and bedrock, into the aquifer that fed one of the town's wells. The public inquiry into the tragedy found that the farmer had used provincially approved methods of spreading manure and could not be faulted.[2] Fault aside, cattle manure was a critical component of the tragedy – a fact that called widespread attention to the severity of the risks posed by farms.

The Walkerton tragedy changed the way many Canadians think about agriculture. But there is nothing new in manure contaminating surface water or groundwater, threatening the health of those who drink the water, swim in it, or eat shellfish grown in it. Because pathogens, such as E. coli and cryptosporidium, are common in manure – one U.S. study found a toxic strain of E. coli in cattle manure at 72 per cent of feedlots[3] – and because they can live for weeks or months, they frequently end up in our water supplies. More than one third of the rural wells tested in several provinces in the 1970s, 1980s, and 1990s exceeded guidelines for coliform bacteria. A major survey of rural wells in Ontario found that 34 per cent exceeded such guidelines. Although septic tanks doubtless contributed to the contamination, feedlots, exercise yards, and agricultural fields were identified as primary culprits. Smaller surveys of wells in other provinces have found comparable levels of contamination: 26 to 27 per cent in Quebec; 21 to 37 per cent in New Brunswick; and 37 to 43 per cent in Manitoba.[4]

Although well water can be treated[5] or avoided, and although rural residents develop immunity to some pathogens, contaminated wells do pose public health problems.[6] Living in a livestock-intensive region can increase the risk of gastrointestinal illness. In the early 1990s, a Health Canada scientist mapped 3,000 reported E. coli cases in Ontario and found a strong and persistent association with cattle density.[7] Alberta's highest rates of intestinal disease are found in 'Feedlot Alley,' where hundreds of livestock operations produce millions of tons of manure each year.[8] Cattle waste is suspected of having caused seven outbreaks of water-borne disease in British Columbia in the last decade, including those involving salmonella, giardia, camplyobacteria, and cryptosporidium.[9] In its 2004 report on Great Lakes Water Quality, the International Joint Commission expressed its ongoing concerns about microbial pollution from farms and other sources, warning that 'the potential for tragedy remains if drinking water is ... challenged by high pollution loads.'[10]

While manure pollution is not new, the scale upon which manure is being produced, stored, and spread on fields *is* new. Recent decades have witnessed a transformation in agriculture. The number of farms has fallen, and those remaining have intensified their operations. Larger, more specialized farms – often referred to as factory farms – now dominate the industry. The trend has been most pronounced on pig farms: while the number of such farms declined to 15,472 from 55,765 between 1981 and 2001, the average number of pigs per farm increased to 902 from 177.[11] It is no longer unusual for pig farms or cattle feedlots to house 10,000 or 20,000 animals. One firm recently proposed – unsuccessfully – an Alberta facility that would have produced 150,000 pigs a year.

As herds have grown, so, too, have barns and manure storage facilities. Machinery has become more powerful; and the use of inputs such as water, fuel, fertilizers, and pesticides has increased.[12] Expanding operations have provided some farmers with the capital, technology, and expertise to ensure the sustainability of their operations. But intensification has often increased risks to the environment and to human health and well being.[13] In many intensively farmed areas, public appreciation of the environmental benefits of traditional farming – the preservation of bucolic landscape, the protection of groundwater, the provision of wildlife habitat – has given way to concerns about modern agriculture's impacts on water, soil, air, and human health. Such concerns, once primarily local, have reached federal environmental authorities. In 2003, Environment Canada pronounced itself 'very concerned about the potential negative environmental effects of agricultural operations.'[14]

Threats posed by manure dominate concerns about the industry. A 2006 survey recorded almost 15 million cattle and more than 14 million pigs on Canadian farms.[15] The livestock generate enormous amounts of feces and urine. Canada's cattle, hogs, poultry, horses, sheep, and goats produced an estimated 177.5 million tonnes of manure in 2001 – almost 22 million tonnes more than they produced in 1981.[16] In Ontario and Quebec alone, the manure generated by livestock equals the sewage produced by more than 100 million people.[17]

Many farmers manage their manure irresponsibly, shunning effective treatment technologies and sustainable operating methods. In the words of one American critic, 'Even though the industry uses state of the art methods for meat production, it is still using "slop bucket" technology for waste treatment.'[18] Most Canadian dairy and hog farmers do not treat their manure. Uncovered manure storage systems remain common. Many are poorly situated: thousands are within thirty metres of a well or are near a stream, river, or lake.[19] Manure spills occur frequently; Ontario farms reported 274 between 1988 and 1998.[20] Some spills involve large volumes of manure; in 2002, almost one million gallons of manure spilled from a storage tank on a Manitoba hog farm.[21]

Many manure storage systems are too small, forcing farmers to spread their manure on nearby fields in unsuitable conditions. Almost one fifth of Canada's hog farmers spread manure during the winter, when it is more likely to run off frozen ground. Whatever the season, once spread, manure is too often allowed to sit on the soil's surface. More than half of Canada's livestock farmers either leave manure on the surface or incorporate it into the soil more than a week after applying it to the land.[22]

Poorly managed manure can wreak environmental havoc. In addition to contaminating waters with E. coli and other pathogens, manure can contaminate waters, ground and surface alike, with nitrogen and phosphorus. Nitrogen and phosphorus are valuable plant nutrients, sustaining the health and growth of crops when applied to fields in the right quantities and at the right times. But when manure run-off from

spreading, spills, or leaks reaches lakes or rivers, these nutrients 'fertil-
ize' the water, spurring algae and other aquatic plant growth, reducing
the oxygen in the water and killing fish. In high concentrations, ammonia, a form of nitrogen, kills fish directly. Nitrate, another form of nitrogen, kills frogs and toads more slowly.

Nutrients from manure and fertilizer threaten lakes and rivers across
Canada, particularly in areas of intensive agriculture. More than two
thirds of the farmland in the humid region of BC poses high risks of
water contamination. Nutrient levels in Alberta's streams and lakes
regularly exceed provincial guidelines: in samples taken from streams
in intensively farmed areas, 99 per cent exceeded phosphorus guidelines, while 87 per cent exceeded nitrogen guidelines. Phosphorus levels likewise exceeded guidelines in most rivers in agricultural areas of
southwestern Quebec.[23] Canada's Commissioner of the Environment
and Sustainable Development reported in 2001 that the misuse of
manure and fertilizers has damaged the ecosystem of the Great Lakes
and St Lawrence River Basin. She warned that nitrogen is accumulating
in agricultural soils in both Quebec and Ontario and that 'nutrient
problems are getting worse.'[24] And in Manitoba, the run-off of nutrients from farmers' fields is contributing to the degradation of Lake
Winnipeg – a lake that ecology Professor David Schindler calls 'Canada's sickest body of water.'[25]

Nutrients likewise pollute groundwater. Nitrate, highly soluble, can
leach through the soil and contaminate drinking water in wells, increasing the risks of kidney and spleen problems and bladder cancer. As
well, nitrate's interference with the blood's ability to deliver oxygen can
cause the potentially fatal 'blue baby syndrome' in infants – a threat that
prompted Ontario's Ministry of Health to warn that no infant under the
age of six months should be fed formula made with well water.[26] Agriculture and Agri-Food Canada calls nitrate levels in groundwater 'a
continuing concern.'[27] Unacceptable concentrations have been found in
virtually every province. Typically, between 5 and 20 per cent of wells
are contaminated with unsafe levels of nitrate. Some regional surveys
have reported excess concentrations in up to 64 per cent of the wells
surveyed.[28]

Manure threatens not only the water we drink but also the air we
breathe. 'I stink therefore I ham,' says one cartoon pig philosopher to
another.[29] And indeed, livestock farming is an inherently smelly business. Whether accumulating in barns, decomposing in open lagoons, or
being sprayed on fields, animal wastes inevitably release noxious

gases, several of which create extremely unpleasant odours.[30] The odours produced by livestock operations frequently plague nearby residents; in fact, they comprise the largest source of complaints from neighbours to municipal officials.[31] While it has long been understood that livestock odours annoy neighbours and diminish the quality of their lives, only recently have scientists begun to look into the adverse health effects of these odours. A groundbreaking study, published in 1995, found that neighbours exposed to swine odours reported elevated levels of tension, depression, anger, fatigue, and confusion.[32] In the intervening decade, a growing body of research suggests that exposure to livestock odours, and the resulting psychological and physiological stress, can cause physical illness.[33]

Some odorous emissions from farms are themselves poisonous. Others simply serve as 'markers' for non-odorous emissions that can threaten human health. The air in barns is often contaminated with toxic gases (such as ammonia and hydrogen sulfide), dust (from feed, animals, or dried feces), and bioaerosols (particles that contain endotoxins, bacteria, and fungi). There is no question that such contaminants, alone or in combination with one another, can make people ill.[34] Since the 1970s, dozens of studies have documented respiratory problems such as bronchitis, asthma, sinusitis, and throat irritation among swine and poultry workers.[35] With the intensification of livestock operations and the accompanying rise in emissions from barns, manure storage facilities, and the manure spread on fields, scientists have begun to study whether people living near farms suffer similar health problems. Limited studies to date have found elevated rates of headaches, runny noses, sore throats, coughing, asthma, burning eyes, plugged ears, weakness, nausea, and diarrhea among intensive farms' neighbours.[36] One study of the discharge records of a hospital near one of the United States's largest hog operations found that diarrheal cases quadrupled and cases involving respiratory illness tripled when the facility became operational.[37]

Although much remains to be learned about the severity and reach of agriculture's threats to human health, concern is mounting.[38] In 1998, experts brought together by the U.S. Centers for Disease Control agreed that 'adequate evidence currently exists to indicate airborne emissions from large-scale swine facilities constitute a public health problem.'[39] In 2002, concerns about the public health risks posed by industrial hog farms prompted the Canadian Medical Association to appeal for a moratorium on the expansion of the industry pending further research. A year later, the American Public Health Association followed suit, calling for a moratorium on new concentrated animal feed operations until more is known about the risks they pose to public health.[40]

Concerns are growing about several other links between animal production and human health. Manure contains antibiotics, the widespread use and environmental dispersion of which may favour the proliferation of antibiotic-resistant bacteria.[41] Bacteria resistant to tetracycline surfaced in hogs in the UK in 1956. Since then, antibiotic resistance has appeared in many of the pathogenic bacteria commonly associated with livestock, such as E. coli and salmonella.[42] Antibiotic-resistant bacteria may be transmitted from animals to humans through exposure to infected animals or consumption of contaminated food.[43] According to the International Joint Commission, 'some experts believe that the massive and largely unregulated use of antibiotics in agriculture and aquaculture, coupled with the increasing number of antibiotic-resistant pathogens found in nature, may present the greatest risk to the aquatic environment and to public health.'[44] The Canadian Committee on Antibiotic Resistance, estimating that antibiotic resistance now costs the Canadian health system $700 million annually, has warned that annual costs could rise to $1.8 billion if the prevalence rises to U.S. levels.[45]

Manure is by no means the only agricultural pollutant, and intensive livestock operations are by no means the only farms that affect the environment. Agriculture has a number of other environmental impacts, some intensely local, others affecting people far downstream or downwind. Almost three quarters of Canada's farms apply herbicides, insecticides, or fungicides to their crops.[46] Many pesticides release smog-forming gases into the air. Eroded soils carry pesticides to surface waters; pesticides also reach waters through precipitation or atmospheric deposition. Pesticides contaminate groundwater by leaching or, more often, by flowing through cracks in the soil, insect or animal burrows, or holes left by decaying roots. Pesticides are frequently detected in wells, farm dugouts, ponds, lakes, and streams. Although concentrations do not generally exceed Canada's relatively lax guidelines for drinking water quality, they more often exceed guidelines for irrigation water and for the protection of aquatic life.[47] Pesticides are of particular concern in Prince Edward Island, where run-off from potato fields caused twenty-three fish kills in seventeen rivers between 1994 and 2002.[48]

The adverse effects of pesticides are fiercely debated. Concerns about the natural environment focus on the effects of pesticides on fish, amphibians, and birds. Myriad studies have raised red flags about the effects of pesticides on human health – particularly on farm children's health. While acknowledging the need for more research, the Ontario College of Family Physicians has warned that the data 'tend towards implicating pesticides as inducing damage to children's immune, endocrine, nervous and reproductive systems, as well as congenital anomalies and cancer.'[49] Others have reported links between pesticides and childhood leukemia, kidney tumours, brain tumours, and learning and memory problems.[50] Still others have suggested links with breast cancer, Parkinson's disease, prostate cancer, and low sperm counts.[51]

Farming, conducted irresponsibly, can adversely affect the environment in many other ways, as well. Some farming practices cause erosion. Sediments from farmers' fields damage fish and other aquatic life. They accumulate in municipal drains, ditches, rivers, lakes, and harbours, interfering with navigation and necessitating costly dredging. They increase the costs of water treatment. Although precise data are unavailable, an estimated 650,000 tonnes of sediment are deposited in the Great Lakes every year, costing Ontario well over $100 million annually.[52]

And the list goes on. Some agricultural operations consume considerable quantities of water.[53] Some generate loud noises, either inadvertently, through the use of machinery, or intentionally, in order to scare birds and wildlife. Some destroy wetlands and wildlife habitat. Some harbour viruses, such as the avian flu, that can threaten animals and humans alike. Some generate dust. Some, in emitting ammonia that mixes with other pollutants, create a milky white haze, described by Environment Canada as 'a rural version of urban smog.'[54] And some create potent greenhouse gases. Indeed, agriculture is thought to be responsible for one quarter of Canada's methane emissions, more than one half of the country's nitrous oxide emissions, and about one tenth of the country's total greenhouse gas emissions.[55]

Unfortunately, Canadians lack comprehensive information about many of the environmental effects of agriculture. Monitoring programs tend to be woefully inadequate, providing only localized data for limited periods of time.[56] Even so, the information that is available paints a picture of an industry whose impacts are serious, widespread, and worsening. None of these impacts can remain unchecked if farming is to be sustainable. Controls are clearly required. But what kinds of controls do agriculture's various impacts call for? The controls currently in place appear to be ineffective. Do other regulatory authorities using other instruments hold greater promise? Who should regulate what? And what tools will they require? These are the questions that this book sets out to answer.

The following chapters chronicle a gradual centralization of control over agricultural pollution, starting in the 1970s and continuing to this day. Traditionally, the control of farms' impacts occurred at the local level. The common law enabled individuals to curb nuisances that harmed them. Under that system, which governed agriculture for centuries, farmers' neighbours could obtain injunctions against practices that created unreasonable odours, dust, or noise. In the last century, control increasingly occurred at the municipal level as well. Using their zoning and land-use planning authority, municipalities limited the types and sizes of farms permitted within their boundaries.

Both individual and municipal control effectively restrained the unsustainable intensification of agriculture – *too* effectively, in fact, for provincial governments, which responded by transferring most decision-making authority to themselves and their appointees. Through 'right-to-farm' legislation, all provinces, regardless of the party in power, have curtailed rural residents' common-law rights to be free of agricultural pollution. Many have likewise disempowered rural municipalities, insisting that they heed provincial policies. Because such policies have almost universally favoured agriculture, the wresting of control from individuals and communities has had the effect of permitting levels of agricultural pollution that were once widely forbidden.

Although the provinces currently dominate agricultural regulation, signs point to growing federal involvement. Federal officials are becoming more critical of agricultural pollution. Environment Canada, the Department of Fisheries and Oceans, and provincial environment ministries are more often enforcing federal Fisheries Act prohibitions against surface water pollution from farms. Liberal David Anderson, when he was federal environment minister, mused about the need for national mechanisms to protect groundwater from the risks posed by intensive livestock operations.[57] NDP leader Jack Layton has called on the federal government to take responsibility for BC's aquaculture industry; in so doing, he drew a parallel with the issue of intensive hog barns, implying that he would like to see a federal role in regulating the latter, as well.[58] At least one prominent environmental group has called on the federal government to establish wide-ranging health and environmental standards for intensive farms.[59] And a cross-Canada coalition organized by the Council of Canadians seeks a national Clean Water Act in its quest for sustainable livestock production.

Increasingly, however, problems with government control of agricultural pollution are emerging. Too often, governments are being asked to control pollution that they themselves have created by subsidizing unsustainable farming practices. And too often, politics trump science in decisions about agriculture. Agriculture Canada documents from the late 1990s indicate that the agency promoted only 'good news' stories about agriculture while suppressing research indicating that the sector posed significant health hazards and environmental problems. Alarmed by the department's bias, University of Brandon biology professor Bill Paton warned, 'People think the government is looking after them and we're finding out [it's] not.'[60]

Nor can provincial agencies be counted on to look after the public interest. Agricultural ministries are often expected to both promote and regulate their industry. Ontario's experience is illustrative. The report of the Walkerton Inquiry called attention to the potential conflict of interest within the Ontario Ministry of Agriculture, Food, and Rural Affairs, pointing out that the ministry 'could be seen to be simultaneously promoting the needs of the agricultural community and regulating that community. The possibility of such a perception has increased in the past few years, during which time OMAFRA has focused strongly on rural economic development and provided less attention to environmental protection.'[61]

Successive provincial environmental commissioners have also noted that OMAFRA's allegiance to farmers may compromise its regulatory effectiveness. In 1998, OMAFRA removed from its Statement of Environmental Values the commitment to 'ensure an environmentally responsible and sustainable agriculture and food system.'[62] The ministry's 'vision' is now 'to foster competitive, economically diverse and prosperous agricultural and food sectors and promote the economic development of rural communities.' As one environmental commissioner warned, since OMAFRA staff tend to handle farm nuisances, 'farm discharges may not be dealt with as vigorously as industrial discharges and emissions.'[63]

Financial constraints may also impede enforcement.[64] The Rural Rights Alliance of Ontario, which has complained that the environment ministry ignores agricultural run-off and leaking lagoons, has blamed a lack of funding: 'With staff and resources slashed, they, along with the

Ministry of Natural Resources, have chosen not to respond when contacted about countless spills. Likewise, local health units have turned a blind eye to direct continual threats to well water and public safety.' But the alliance has also blamed politics, complaining that OMAFRA has led opposition to the strengthening of laws prohibiting agricultural pollution. 'Our experiences with OMAFRA,' it laments, 'have been farcical.'[65]

Whatever the reasons, political or financial, Canada's provincial and federal governments are not writing effective laws and regulations and they are not enforcing those that exist. They consistently choose to influence farmers' behaviour through education and moral suasion rather than strict regulation. In 2002, defending the federal government's approach to regulation as embodied in its promotion of voluntary 'environmental farm plans,' then federal agriculture minister Lyle Vanclief maintained, 'Our Canadian farmers don't need a regulatory framework; nobody wants [environmental farm plans] as much as they do.'[66] The minister seemed oblivious to the pointed criticism issued by the federal environment commissioner the previous year: 'No link has been made between the environmental farm plans and observable benefits to the environment, such as better water quality. The programs give farmers little specific incentive to minimize their impacts on the environment beyond the farm gate.'[67] In 2005, the commissioner renewed her criticism of federal regulators, charging that 'Environment Canada cannot yet demonstrate that its compliance promotion and enforcement efforts at hog farms are effective' and that 'Agriculture and Agri-Food Canada's strategic approach to reducing the environmental impacts of hog farming is not clear.'[68]

What *is* increasingly clear to many environmentalists and rural residents is that they cannot wait for provincial or federal governments to prevent or halt agricultural pollution. As one frustrated Ontarian announced at a public meeting about water quality, 'You have to realize that they're not doing much ... you pretty well have to do it on your own.'[69] Doing it on one's own, however, is often no longer possible, thanks to the disempowering of citizens that is documented in the following chapters. Governments' systematic overriding of individuals' common-law property rights has left rural residents at the mercy of ever-intensifying farms, helpless witnesses to the growing threats to their comfort, their health, and their environment.

In documenting the environmental and social costs associated with the centralization of decision making, this book suggests caution regarding further centralizing measures. It proposes a regulatory regime that respects the principle of subsidiarity, placing decision making as close as possible to affected citizens. Restoring control of agriculture's impacts to the lowest workable level will best reflect the unique needs and values of affected individuals and communities. Only when local environmental impacts are addressed locally and broader impacts are regulated at higher levels will farming become truly sustainable.

2 Severing the Gold from the Dross: Using the Common Law to Curb Unsustainable Farming Practices

Pollution is always unlawful and, in itself, constitutes a nuisance.
– Supreme Court of Canada Justice Thibaudeau Rinfret, 1928[1]

Conflicts about farming are nothing new. For centuries, farmers and their neighbours have argued about livestock containment, manure storage and application, water management, and other practices affecting those living downwind or downstream of farms. Until recently, such conflicts were usually resolved by the parties directly involved – through discussion and negotiation or, when that failed, with the aid of the common-law courts. In pre-industrial England, where the common law developed, farmers' neighbours frequently called upon local courts to adjudicate disputes about straying cattle, odorous pigsties, and polluted streams. Drawing on local norms and values, early courts developed principles governing the uses of land and water. As the common law evolved through the intervening centuries, judges looked to their predecessors for guidance and followed the precedents set by them. Thus, a number of the principles that emerged hundreds of years ago continue to guide common-law decisions to this day.

The age-old common-law maxim, 'use your own property so as not to harm another's,' has provided the foundation for the resolution of disputes about farming practices, balancing the interests of farmers with those of their neighbours. Under this maxim, farmers' rights – like those of others who own or occupy land – are tempered by their responsibilities. While farmers have a right to use and enjoy their property, they have a responsibility not to interfere with their *neighbours'* rights to use and enjoy *their* property.

This fundamental principle of the common law was articulated in England as early as the thirteenth century, when one legal scholar wrote that 'no one may do in his own estate any thing whereby damage or nuisance may happen to his neighbour.'[2] Courts frequently cited the principle in the following centuries. In 1611, an English court hearing a dispute over the erection of a pigsty that corrupted the air relied on the maxim in its original Latin: *sic utere tuo ut alienum non laedas*.[3] A 1703 court decision included the following discussion of the principle:

> Every man is bound so to look to his cattle, as to keep them out of his neighbour's ground, that so he may receive no damage ... So if a man has two pieces of pasture, which lie open to one another, and sells one piece, the vendee must keep in his cattle so as they shall not trespass upon the vendor. So a man shall not lay his dung so high as to damage his neighbour, and the reason of these cases is, because every man must so use his own, as not to damnify another.[4]

In the late eighteenth century, 'use your own property so as not to harm another's' was described as 'the rule' by the famous jurist, Sir William Blackstone, in his *Commentaries on the Laws of England*.[5] Court decisions of that era confirm that judges applied the rule without question or hesitation.

Canadian courts likewise embraced the principle. In an 1896 decision concerning a Montreal stable that produced offensive odours, the Chief Justice of the Supreme Court of Canada noted that the maxim was 'as much a rule of the French law of the province of Quebec as of the common law of England.' He explained that even the most absolute proprietary rights 'must, according to the general principles of all systems of law, be subject to certain restrictions subordinating the exercise of acts of ownership to the rights of neighbouring proprietors.'[6] The venerable principle continued to guide courts throughout the twentieth century. In a 1967 decision concerning a hen-laying business in Ontario, the judge relied on a well-known legal text identifying *sic utere tuo ut alienum non laedas* as 'the basis of the law of nuisance.'[7]

This central tenet of the common law has provided an extremely effective barrier – indeed, a virtual wall – against agricultural pollution. As an English court explained in a 1711 decision concerning straying cattle, 'the law bounds every man's property, and is his fence.'[8] Blackstone expanded upon the notion of the law as a fence: 'Every man's land is in the eye of the law enclosed and set apart from his neighbour's: and that either by a visible and material fence, as one field is divided from another by a hedge; or by an ideal invisible boundary, existing only in the contemplation of the law.'[9]

A farming activity that breaches common-law boundaries tends to fall into one of three categories: it may constitute a trespass, a nuisance, or a violation of someone's riparian rights.[10] Under the common law, it is a trespass to place anything – be it large or small, toxic or harmless – upon someone else's property. Blackstone defined trespass very broadly as 'any entry on the property of another.' The law's application to farming was clear. 'A man,' Blackstone explained, 'is answerable for not only his own trespass, but that of his cattle also: for, if by his negligent keeping they stray upon the land of another, and much more if he permits, or drives them on, and they there tread down his neighbour's herbage, and spoil his corn or his trees, this is a trespass, for which the owner must answer in damages.'[11] A farmer may also be answerable under trespass for the direct invasion of another's land by tangible pollutants, such as manure or dust.

For indirect or intangible invasions, such as odours and noises, the common law of private nuisance usually applies. As discussed in greater detail below, a nuisance is something that unreasonably interferes with another's use or enjoyment of his property. Nuisance involves an element of harm. Indeed, the term comes from the French *nuire* and, in turn, from the Latin *nocere* – both meaning 'to do harm.' Blackstone defined nuisance as 'anything that works hurt, inconvenience, or damage.' Again, he used farming to illustrate the tort: 'If a person keeps his hogs, or other noisome animals, or allows filth to accumulate on his premises, so near the house of another, that the stench incommodes him and makes the air unwholesome, this is an injurious nuisance, as it tends to deprive him of the use and benefit of his house.'[12] Blackstone's catalogue of nuisances also included 'destroying one's peace and comfort by unusual noises' and corrupting or poisoning a watercourse.

More often – at least, in England, eastern Canada, and the eastern United States – common-law disputes about water pollution have fallen under the heading of riparian rights. Under the common law, riparians – those who own or occupy land beside lakes and rivers – have the right to the natural flow of water beside or through their property, unchanged in quantity or quality. In 1900, one judge summarized the law in this way: 'Every riparian proprietor is entitled to have the waters of the stream that washes his land come to it without obstruction, diversion, or corruption.'[13]

Once a trespass, nuisance, or violation of riparian rights has been established, a court will award damages and/or issue an injunction – an order requiring the cessation of the offensive activity or, in some cases, specifying corrective action. Damages are important in that, if appropriately set, they compensate the plaintiff for his injury.[14] However, damages alone cannot ensure that the injury will not continue. For this reason, common-law courts were at one time required to issue injunctions. Although they now have the authority to substitute damages for injunctions, they are generally reluctant to do so.

The most influential case on this matter was decided in 1894. Called *Shelfer v. City of London Electric Lighting Company*, it dealt not with a nuisance created on a farm but with steam, noise, and vibrations caused by electricity generators. In his decision, Lord Justice Lindley stressed that a court should award damages instead of an injunction only 'under very exceptional circumstances.' The court, he insisted, 'has always protested against the notion that it ought to allow a wrong to continue simply because the wrongdoer is able and willing to pay for the injury he may inflict.'[15]

His colleague, Lord Justice Smith, agreed. 'A person committing a wrongful act,' he wrote, 'is not thereby entitled to ask the court to sanction his doing so by purchasing his neighbour's rights, by assessing damages in that behalf, leaving his neighbour with the nuisance.' He suggested that damages should be substituted for an injunction only in the following circumstances: if the injury to the plaintiff's legal rights is small; if the injury is capable of being estimated in money and can be compensated by a small payment; and if the case is one in which it would be oppressive to the defendant to grant an injunction.[16]

The decision in *Shelfer* was extraordinarily influential in the following century. Courts cited it in countless decisions, including that in a 1928 Toronto case concerning a noisy dairy that kept neighbours awake and constituted a nuisance. Although the judges in that case disagreed on the appropriate remedy, they all agreed on the importance of the principles articulated in *Shelfer*. One judge noted that the decision 'has never been questioned in England or here' and that the rule proposed in it 'has been accepted everywhere as a sound guiding principle.' Another judge agreed that the case 'has uniformly been approved and followed' and 'adopted and applied in cases without number.' He approvingly cited the rule that courts' authority to substitute damages for an injunction ought only to be exercised under very exceptional circumstances. Finding no evidence that this rule had been modified in Canada, he concluded that '*prima facie* and in the absence of special circumstances of a sufficiently strong character an injunction should be granted.'[17]

The principles that have guided common-law decisions for centuries – that one must use one's property so as not to harm another's, and that injunctions are the appropriate remedy for all but the smallest violations of this rule – have helped courts address a number of challenges posed by farming operations across Canada. A review of twentieth-century Canadian judgments reveals that courts used these principles as touchstones, applying them to an array of disputes involving farmers and their neighbours, clarifying, elaborating upon, and adapting them as required by novel practices. In so doing, they long ensured that Canadian farmers did little lasting harm to their neighbours or to the environment.

In 1967, an Ontario judge found that smells emanating from henhouses in Tavistock were a nuisance. Six neighbours had complained that the smells – variously described as obnoxious, terrible, sickly, and awful – prevented them from using their gardens and forced them to close their windows. The judge hearing the case cited a well-known legal text's definition of nuisance as 'the act of wrongfully causing or allowing the escape of deleterious things into another person's land – for example, water, smoke, smell, fumes, gas, noise, heat, vibrations, electricity, disease-germs, animals, and vegetation.' An elaboration followed:

> The generic conception involved in nuisance may perhaps be found in the fact that all nuisances are caused by an act or omission, whereby a person is unlawfully annoyed, prejudiced or disturbed in the enjoyment of land; whether by physical damage to the land or by other interference with the enjoyment of the land or with his exercise of an easement, profit or other similar right or with his health, comfort or convenience as occupier of such land ... Thus a judicial definition which has been cited with approval is the following: 'Private nuisances, at least in the vast majority of cases, are interferences for a substantial length of time by owners or occupiers of property with the use or enjoyment of the neighbouring property.'[18]

After determining that the smells constituted a nuisance, the judge turned to the question of the appropriate remedy. The defendant urged him to award damages instead of issuing an injunction, arguing that the plaintiffs could be compensated in money and that moving the henhouses would be a great hardship. The judge called this contention unsound, referring to a contemporary judgment that, in turn, cited the

Shelfer case on the inadequacy of damages: 'No one should be called upon to submit to the inconvenience and annoyance arising from a noxious and sickening odour for a "small money payment," and the inconvenience and annoyance cannot be adequately "estimated in money."'[19] The judge therefore issued an injunction, staying it for six months in order to give the defendant an opportunity to remedy the problem.

In 1977, a Nova Scotia court deemed nuisances the noises from a horse barn outside of Glace Bay, along with the smells generated by the accumulation of manure and the washing of the horses. In grappling with the issue of what constitutes a nuisance, the judge acknowledged his ambivalence, saying that 'the matter is so much a matter of degree.' On this issue he cited a widely used book on the law of torts:

> In nuisance of the third kind, 'the personal inconvenience and interference with one's enjoyment, one's quiet, one's personal freedom, anything that discomposes or injuriously affects the senses or the nerves,' there is no absolute standard to be applied. It is always a question of degree whether the interference with comfort or convenience is sufficiently serious to constitute a nuisance. The acts complained of as constituting the nuisance, such as noise, smells or vibration, will usually be lawful acts which only become wrongful from the circumstances under which they are performed, such as the time, place, extent or the manner of performance.[20]

The judge decided that the number of horses stabled in the barn 'tips the scales in favour of the plaintiffs.' When the defendants increased the number of horses they kept, they 'greatly increased the dimensions of the matters complained of by the plaintiffs. It seems to me that the defendants acted unreasonably in subjecting the plaintiffs to such an increase in objectionable factors associated with the keeping of horses in such an area.'[21]

In 1982, a New Brunswick court determined that two farmers operating a 3000-pig farm close to a resort lake had created a nuisance. In that case, a group of cottage owners had complained of the odours emitted by the farm and of the water contamination caused by the farm's refuse. The trial judge concluded that the farmers had created 'intolerable and dreadful pollution'; they caused a build-up of algae and slime in the lake and emitted a stench that 'can make you ill within a minute.'[22]

In his decision, the judge reviewed a number of recent cases and legal texts spelling out what constitutes a nuisance. He included the following:

The paramount problem in the law of nuisance is, therefore, to strike a tolerable balance between conflicting claims of landowners, each invoking the privilege to exploit the resources and enjoy the amenities of his property without undue subordination to the reciprocal interests of the other. Reconciliation has to be achieved by compromise, and the basis for adjustment is reasonable user ... Reasonableness in this context is a two-sided affair. It is viewed not only from the standpoint of the defendant's convenience, but must also take into account the interest of the surrounding occupiers. It is not enough to ask: Is the defendant using his property in what would be a reasonable manner if he had no neighbour? The question is, Is he using it reasonably, having regard to the fact that he has a neighbour?[23]

A Nova Scotia judge cited the same passage (noting that it had been approved by the Supreme Court of Canada) in a 1990 decision concerning dust from a grain drying and storage facility operated by the provincial grain commission.[24] The judge determined that dust from the road and driveway leading to the facility interfered with a neighbour's enjoyment of her property and that the commission was liable in nuisance.

Most agricultural nuisance cases – including but by no means limited to those discussed above – have dealt with common irritants such as odour, noise, and dust. Farmers' neighbours have also turned to the courts to resolve disputes concerning a variety of other interferences. In one 1972 case, two BC women sued their neighbour and an aerial spraying company when insecticide sprayed on the neighbour's pea crop drifted over their land. The BC Supreme Court concluded that the spraying constituted a nuisance. Although the spray had not harmed the women, it, along with the roar of the plane, had frightened them, as it would have frightened any ordinary person. In nuisance, the Chief Justice noted, 'the burden of proving damage is a relatively easy one in that, where there is an interference by noise or smells with the amenities of living, no permanent loss or injury to health need be proved.'[25]

A 1980 Nova Scotia case concerned contamination of a well by manure spread on neighbouring pasture lands. The plaintiff had 'top dressed' its lands with eighteen tons of manure. It had done so when the lands were heavily frosted, since frost facilitated the movement of the machinery used for spreading. Soon thereafter, an unusually heavy rain washed some of the manure onto the plaintiff's land and into his well, pushing coliform counts to 300 times acceptable limits. The judge found this situation to be 'a classic application of the long established principle' set out in the 1868 British case, *Rylands v. Fletcher*. He cited the rule articulated in that decision: 'If a person brings, or accumulates, on his land anything which, if it should escape, may cause damage to his neighbour, he does so at his peril. If it does escape, and cause damage, he is responsible, however careful he may have been, and whatever precautions he may have taken to prevent the damage.'[26] The rule, he said, continues to apply: 'In my view, the rule in *Rylands and Fletcher* was and still is a very pragmatic and reasonable resolution of the inevitable conflicts and competing interests of adjoining property owners.'[27]

Bees from an apiary in Vernon, British Columbia, were the subject of a 1988 case. When neighbours built a swimming pool, hundreds of water-seeking bees invaded their property, causing, in the judge's words, 'misery and discomfort' and interfering with the neighbours' use of their property. The judge, concluding that the keeping of bees in that location during the summer months was 'obviously a nuisance that could and should be avoided,' awarded damages and issued an injunction.[28]

Nuisance law, inherently flexible, is capable of addressing challenges unimagined during its evolution. In providing a framework of principles, it enables judges to assess an almost endless variety of novel situations. This adaptability to new circumstances has long been a selling point of the common law. In his seventeenth-century commentary on the laws of England, the famous jurist Sir Edward Coke praised the merits of trying any innovation with the rules of the common law, 'for these be true touchstones to sever the pure gold from the dross and sophistications of novelties and new inventions.'[29] And indeed, in modern Canada, common-law rules have served as effective tools to distinguish sustainable agricultural innovations from those that are unsustainable, permitting only the former.

That judges may be asked to apply the most ancient of principles to the most modern of challenges is exemplified by a case now before the Saskatchewan courts. Organic farmers in that province are trying to bring a class-action suit against two biotechnology companies. In 2002, two farmers filed a claim on behalf of all of Saskatchewan's organic grain farmers alleging that genetically modified canola had, through blowing seeds and cross-pollination, contaminated their crops. They complained that, as they could no longer warrant their crops to be GMO-free, they had lost their crops and the markets for them. Among other things, they argued that 'the introduction of GM canola into the

Saskatchewan environment ... created a nuisance that has interfered
with certified organic grain farmers' use and enjoyment of their land.'[30]
The farmers sought not only damages but also an injunction prohibiting the companies from testing genetically modified wheat, which they
feared would lead to the contamination of their wheat crops and the
loss of the markets for these as well. The case has been delayed by questions of whether it can be tried as a class-action suit. Even before it is
decided, however, it illustrates the constantly evolving applications of
the common law.

In responding to charges of nuisance, farmers have proposed countless defences. Courts have generally rejected a number of farmers'
more common excuses – that their actions were necessary, that they had
taken care to act properly, that they had been carrying on the disputed
activity before the plaintiff moved to the area, or that a judgment
against them would create too great a burden.[31]

The defence of necessity has rarely persuaded courts. As early as
1611, one William Alfred tried to excuse his smelly pigsty on the
grounds that 'the building of the house for hogs was necessary for the
sustenance of man.'[32] The court rejected this argument, and its succes-
sors have, by and large, continued to do so. Addressing this issue, two
of the above-mentioned contemporary judgments looked back more
than 100 years to an English decision that they could not improve upon:
'It is no defence to an action for nuisance to show that the defendant's
operation of his farm is a useful one necessary to the public interest.'[33]

The defence that a polluting farmer has taken care or exercised caution has also failed to move courts. Aside from the exception discussed
below – farmers who create smoke when clearing agricultural land are
not liable for nuisance if they are not negligent and if the fires reflect
normal practices – Canadian courts have been reluctant to link negligence to nuisance.[34] In the Tavistock hen-house case, the defendant
argued that he had taken all reasonable care to reduce odours and
should therefore not be held responsible. Although the judge credited
him for purchasing expensive equipment, he insisted that taking care
could not exonerate him. To bolster his conclusion, he cited a legal text:
'It is no defence that all possible care and skill are being used to prevent
the operation complained of from amounting to a nuisance ... If an
operation cannot by any care or skill be prevented from causing a nuisance, it cannot lawfully be undertaken at all, except with the consent of
those injured by it or by the authority of a statute.'[35]

Variations on this theme appeared in the judgment in the New Brunswick dispute between the hog farmers and the cottage owners. The judge cited legal decisions from 1913 and 1915 respectively: 'It is no defence to an action for nuisance to show that ... it is carried on with all care and skill and every effort is made to prevent it from being a nuisance ... Their duty to their neighbour is not merely to take care so as to avoid causing a nuisance. Their duty is to abstain from causing one at all.'[36]

In the Nova Scotia case regarding contamination of a well by manure washed from neighbouring pasture lands, the defendant argued unsuccessfully that since top dressing of land is 'an act of normal husbandry,' he should be liable only if proved negligent. In response, the judge cited *Rylands v. Fletcher*: 'In considering whether a Defendant is liable to a Plaintiff for damage which the Plaintiff may have sustained, the question in general is not whether the Defendant has acted with due care and caution, but whether his acts have occasioned the damage.'[37] The judge commented, 'I know of no case in Nova Scotia where the application of the rule in *Rylands and Fletcher* has been in any way modified or eroded for considerations of normal agricultural husbandry.'[38]

Nor can farmers who create nuisances defend themselves on the grounds that they had been carrying on their activities before the plaintiffs moved nearby. While their counterparts in the United States may, in increasingly rare cases, escape liability by demonstrating that their neighbours 'came to the nuisance,' Canadian farmers – like other Canadian polluters – cannot avail themselves of this defence.[39] They are bound by a rule stated by an English law lord in the late nineteenth century: 'Whether the man went to the nuisance or the nuisance came to the man, the rights are the same.'[40] The judge in the Glace Bay horse barn dispute disposed of the defence that the plaintiff had come to the nuisance with the following: 'The law is clear that if a man purchases property in respect of which a nuisance is being committed it is still actionable by him, for it is a nuisance of which he can legally complain.'[41]

The Canadian courts' rejection of the coming to the nuisance defence has drawn criticism. Farmers argue that neighbours who have come to established nuisances likely knew about them in advance and, in choosing to move regardless, implicitly consented to them. They also claim moral authority to create nuisances on the grounds that they were there first, and that first in time makes first in right. Such arguments overlook the fact that farmers were generally *not* there first. Today's farmers did not homestead unowned lands. Lands neighbouring theirs have been owned, whether or not the owners have complained about odours. or other nuisances. Giving farmers the right to continue past nuisances would, in effect, be giving them easements over these neighbouring lands. The easements would reduce the number of uses to which the lands could be put, thereby devaluing the lands. In effect, a coming to the nuisance defence would allow farmers to expropriate portions of their neighbours' properties without compensation.

Accordingly, farmers who have polluted without penalty in the past have not acquired any *right* to pollute or to continue polluting in the future. Although they have had a right to use their own property, they have not obtained any entitlement to use neighbouring lands as receptacles for their wastes – not unless they purchased the affected lands, or purchased easements, or signed contracts with the lands' owners. Those fortunate enough to benefit from the free use of neighbouring lands have merely enjoyed a temporary windfall.[42]

Those who argue for the protection of long-standing nuisances also overlook the extent to which the coming to the nuisance defence would exacerbate agricultural pollution. By providing protection from lawsuits, it would create a moral hazard, diminishing farmers' incentives to reduce the impacts of their operations on the surrounding environment. Worse, the defence would create incentives to establish nuisances. A farmer's land is more valuable with nuisance easements, since they increase the number of uses to which it may be put. A farmer thus has every incentive to grab as many free easements as he can while he can – an incentive that may lead to early investment or overinvestment in polluting activities.[43] Yet another argument against the defence is that it works against progress, entrenching land uses that may be of lower value than the alternatives. The Canadian courts' refusal to permit polluting activities that pre-date a plaintiff's arrival on the scene thus has much to recommend it, both economically and ecologically.

Courts have also consistently rejected the defence that a finding of liability would irreparably damage the defendants' financial interests.[44] The New Brunswick hog farmers who were taken to court by local cottage owners argued that it would be impossible to stop creating odours without disposing of their entire herd. Selling at a sacrifice price, they warned, would cost them more than $60,000 each, thus necessitating personal bankruptcy. The Court of Appeal rejected this defence with the following comments: 'Insofar as it is regrettable that it be this way, these allegations are not relevant to the resolution of the issues before this court. The right to a clean and healthy environment cannot give way to financial interests in a society conscious of each individual's interests. Furthermore, the time to consider this aspect of the issue was before the construction of the pig farm when it was known that this lake was a holiday resort which already had about seventy-five cottages.'[45]

By no means, however, do common-law courts always reject the defences put forward by farmers, nor find the adverse impacts of farms to be nuisances. Indeed, many a court, in balancing the rights of a farmer and his neighbour, has found in favour of the former. While guided by broad principles, courts always consider the specific persons involved in a conflict and the environment in which the conflict occurs. Mitigating circumstances may include the severity of the impacts complained of, the sensitivity of the plaintiff, the reasonableness of the activity, the character of the neighbourhood in which it occurs, and whether or not the government has authorized the activity.

Courts tend to refrain from ruling on trifling amounts of pollution. Generally, the rule of 'give and take, live and let live' governs minor inconveniences or temporary irritants. This rule was discussed in the dispute over the noises and smells of the horse barn near Glace Bay. The judge sought guidance from a book on torts:

> In organised society everyone must put up with a certain amount of discomfort and annoyance from the legitimate activities of his neighbours, and in attempting to fix the standard of tolerance, the vague maxim *sic utere tuo, ut alienum non laedas* has been constantly invoked. As has been pointed out, 'the homely phrases "Give and take" and "Live and let live" are much nearer the truth than the Latin maxim.'[46]

The judge in the Glace Bay case – where one of the plaintiffs suffered from asthma – also explained that unusually sensitive people do not have claims against their neighbours: 'An interference with something of abnormal sensitiveness does not of itself constitute nuisance.'[47] The plaintiff who suffered from asthma was found to be abnormally sensitive; the defendants were not liable for his adverse reactions to their horses. The other plaintiff, however, was found to have a legitimate claim in nuisance.

Likewise, in the Nova Scotia grain commission case, the judge considered the plaintiff's super-sensitivity to dust, referring to a treatise on torts:

> The standard for deciding whether a particular use of land exposes others to an unreasonable interference is objective, in the sense that it has regard to the reactions of normal persons in the particular locality, not to the idiosyncrasies of the particular plaintiff. The law does not indulge mere deli-

cacy or fastidiousness. Thus, it has never been doubted that the question whether smoke, dust or offensive odours cause a sufficient personal discomfort to neighbouring occupiers depends upon its effect on a normal person of ordinary habits and sensibilities.[48]

The judge decided that the plaintiff's use and enjoyment of her property had been impaired quite apart from her sensitivity to dust, and awarded damages for the annoyance and discomfort she had experienced. He noted, however, that most of the damages that she claimed – such as drugs and moving expenses – were related only to her sensitivity. These costs, he determined, were not recoverable.[49]

Before determining liability, courts must assess the reasonableness of a disputed activity. No clear definition of reasonableness exists, as it varies with every circumstance. As explained in one legal text cited in a farming conflict, 'In determining the question whether a nuisance has been caused, a just balance must be struck between the right of the defendant to use his property for his own lawful enjoyment and the right of the plaintiff to the undisturbed enjoyment of his property. No precise or universal formula is possible, but a useful test is what is reasonable according to ordinary usages of mankind living in a particular society.'[50]

In a 1992 decision in which a Manitoba hog farm was found *not* to be a nuisance, the judge cited a 1979 decision by the British Columbia Court of Appeal:

> The test then is, has the defendant's use of this land interfered with the use and enjoyment of the plaintiffs' land and is that interference unreasonable? ... What may be reasonable at one time or place may be completely unreasonable at another ... It is impossible to lay down precise and detailed standards but the invasion must be substantial and serious and of such a nature that it is clear according to the accepted concepts of the day that it should be an actionable wrong. It has been said that Canadian judges have adopted the words of Knight Bruce, V.C., in *Walter v. Selfe* (1851) ... to the effect that actionability will result from an interference with '... the ordinary comfort physically of human existence, not merely according to elegant or dainty modes and habits of living, but according to plain and sober ... notions.'[51]

Although the concept of reasonableness gives courts leeway, most reject the defence if an activity has substantially interfered with another's ordinary use or enjoyment of his land. As explained in one torts text, 'No use of property is reasonable which causes substantial discomfort to others or is a source of damage to their property.'[52] The judge in the Nova Scotia grain commission case adopted just such a position. Citing an earlier Newfoundland decision, he wrote, 'if I use my land then in such a way as to cause injury to my neighbours' land or naturally interfere with his enjoyment of it then this use is unreasonable.'[53]

Even activities that adversely affect farmers' neighbours may not be deemed nuisances if they are in keeping with the character of the neighbourhood in which they occur. If the disputed activity results in material injury or financial harm, the character of the neighbourhood is no defence. But if it results only in personal inconvenience, discomfort, or annoyance, the court will take the character of the neighbourhood into account. Considering the neighbourhood in which a disputed activity occurs allows the court to compromise between 'first in time, first in right' arguments and the general rule, 'use your own property so as not to harm another's.' It pays heed to prior agricultural uses of land but grants them no monopoly. Allowing for gradual change to residential uses, it is evolutionary rather than revolutionary.

In a 1931 decision in which it was determined that a Manitoba fox farm was not a nuisance per se but must take care not to become one, the judge cited several precedents regarding the role of the locality. In one 1865 case, he wrote, it was held that 'what may or may not be denominated a nuisance depends greatly on the circumstances of the place where the things complained of actually occur.' Similarly, he noted, in a 1906 case, the court had found that 'the standard of what amount of freedom from smoke, smell, and noise a man may reasonably expect will vary with the locality in which he dwells.'[54]

In a more recent Manitoba case, the character of the neighbourhood also played an important role in the judge's decision that odours from a hog farm did not constitute a nuisance. The judge noted that the farm was located in an area zoned agricultural. In fact, the local planning statement restricted development to general agriculture and its accessory uses and specified that 'no legitimate farming activity shall be curtailed solely on the basis of objections from nearby acreage owners.' The community had the highest concentration of livestock in Manitoba; the plaintiffs lived within a half-mile of at least five hog operations. The plaintiffs, in short, 'sought to establish an urban type residence in the heart of an intensive hog-producing area.' Although the judge did not doubt that offensive odours had interfered with the plaintiffs' enjoyment of their property, he found that the interference was reasonable in that neighbourhood. He explained, 'There has clearly been interference but it is the unreasonableness of the interference that converts the act from a lawful one into an actionable nuisance ... The Penner operation is consistent with the overall development of the neighbourhood and, by virtue thereof, cannot be considered to give rise to an unreasonable use of the lands in question.'[55]

The British Columbia Supreme Court reached a similar conclusion in 1986 when considering two raspberry farmers' complaint against odours emanating from a neighbouring egg-laying business. The judge dismissed the claim of nuisance, explaining:

> The issue is whether the use of Jaedel Enterprises causes inconvenience beyond what one would expect in this general area. The general area is zoned for agriculture. There are many raspberry farms. There are also poultry, pig and mink farms ... From time to time, throughout the whole year, many pungent odours are expected. Most of these are associated with manure. For short periods, the smells can be overwhelming. Anyone who lives and works in this portion of the country must accept these discomforts. It is an ordinary and normal incident to that locality.[56]

As discussed above, farmers cannot normally avail themselves of the defence of taking care. But one exception has evolved. Farmers who create smoke when clearing agricultural land are not liable for nuisance if the fire is a normal practice and if they are not negligent – if they have conducted themselves 'with such a degree of care and caution as might be looked for in a prudent man.'[57] Why farmers' fires should be treated differently than other nuisances is not clear. Early case law suggests that this exception evolved to reflect fire's necessity and its vulnerability to unforeseeable and uncontrollable changes in wind velocity and direction. A late seventeenth-century British decision mentioned that applying traditional nuisance rules would 'discourage husbandry, it being usual for farmers to burn stubble.'[58] A preconfederation decision in what is now Ontario went further, calling the use of fire to clear land 'indispensable, not merely to individual interest, but to the public good.' The judge implied that the very future of Canada rested on the clearing of land: 'It is not very long since this country was altogether a wilderness as by far the greater part of [it is] still. Till the land is cleared it can produce nothing, and the burning [of] the wood upon the ground is a necessary part of the operation of clearing.'[59]

A 1909 Alberta case concerned damage from a fire that high winds had spread from a nearby farm. The trial judge determined that the defendant farmer was not liable for damages since he had not acted negligently. In his decision, he reviewed in detail the jurisprudence regarding agricultural fires. He explained that the principle governing such fires was rooted in an English decision that established an exception to the general rule of liability as defined in *Rylands v. Fletcher* 'where fire is kindled on land for the ordinary purposes of husbandry, and it escapes without negligence.' Noting that Canadian courts had adopted that rule, the judge referred to an 1861 case in which the court held that 'a person kindling fire upon his own land for the purpose of clearing it, is not liable at all risks for any injurious consequences that may ensue to the property of his neighbours, but is liable only for negligence.' Subsequent courts, he said, had approved that decision 'settled and placed quite beyond dispute' the law as laid down in it.[60]

In a 1973 decision that quoted the above judgment at length, the Appellate Division of the Alberta Supreme Court likewise concluded that a farmer was not liable for two car accidents caused by reduced visibility resulting from smoke drifting from his field onto the roadway. The court set out this test: 'If ... the fire which was started constituted an act of normal husbandry and was conducted without negligence, it would constitute an answer to a claim for nuisance ... The crux of the whole matter depends on whether the acts of Wegner constituted normal husbandry. If they do he is exonerated at least in this province except for negligence.'[61] The evidence indicated that burning stubble and weeds was indeed the usual practice in the area. The court thus determined that 'setting the fire ... was an act of normal husbandry.' Furthermore, the farmer had not been negligent: 'He did everything that any reasonable farmer could be expected to do in relation to the fire.' Accordingly, he could not be held liable for the damage that ensued.[62]

The common-law rule regarding smoke does not, of course, permit farmers to burn with impunity. Negligent practices are strictly forbidden. In a case similar to that described above – again, concerning smoke that reduced visibility and caused a car accident – an Alberta court found negligence on the part of one of the farmers who had, in burning brush, created the smoke. Although the burning 'was done in a normal and common way for this area,' the farmer should have reduced the smoke once he became aware that it was causing a hazard on the neighbouring highway. 'His duty,' the court decided, 'was to stop the continuation of the dangerous situation by controlling its cause ... Having failed to take steps to alleviate the cause of the hazard, he is negligent.'[63]

Overshadowing all other defences is that of statutory authority – the defence that a statute has authorized a disputed activity. Government statutes take precedence over the common law. If a government approves a nuisance, a court loses its power to enjoin it. As an English law lord explained more than a century ago, 'the Legislature is supreme, and if it has enacted that a thing is lawful, such a thing cannot be a fault or an actionable wrong.'[64] At enormous cost to the environment, governments of all times and of all political stripes have overridden the common law to protect favoured industries. Medieval miners, eighteenth-century mill owners, and nineteenth-century railroad companies were among the early beneficiaries of protective legislation. In the last century, Canadian governments, federal and provincial, have aided their own pet industries, including sewage utilities and nuclear power producers, with statutes limiting their common-law liability for pollution.[65] Farmers now benefit from statutes affording some of the country's clearest and most sweeping protections.

In the heyday of the common law, statutory authority rarely appeared in nuisance cases regarding farming. Governments had simply not yet begun to authorize the creation of agricultural nuisances. One exception can be found in a 1901 case, in which a resident of St Thomas, Ontario, brought an action against two railway companies that shipped livestock. She complained that, in feeding hogs in pens near her home, the companies polluted the air with noxious smells and caused her great inconvenience and discomfort. The companies replied that herding large numbers of animals was 'a necessary incident to the carrying on of railway business,' that the required pens would create a nuisance, and that this was contemplated by the Railway Act, the statute under which they operated. The court agreed that the railway companies were 'exercising the powers that they were authorized by law to exercise ... As long as the railway companies have these powers, it is a principle that, exercising them without negligence, they are not liable if, in the proper exercise of the powers, they create a nuisance.'[66]

Today, statutes and regulations govern a wide variety of agricultural activities. A committee established to advise Ontario's agriculture minister in 1986 identified thirty-eight different provincial statutes affecting agricultural planning and operations.[67] But complying with such legislation does not necessarily protect a farmer from common-law liability for nuisance. Courts have honed the defence of statutory authority, limiting its application. In determining whether governments have indemnified particular activities, courts generally distinguish between permissive and mandatory statutes. Under the former, which maintain a farmer's discretion over the choice of operating methods or location, the farmer is expected to act in conformity with private property rights. It is when harm inevitably arises from the discharging of a legislatively imposed duty, or when it is an inevitable consequence of a specifically authorized activity, that the farmer can claim the defence of statutory authority. In mandating an activity or authorizing something to be done in a particular manner or place, the reasoning goes, the legislature sanctions all of its unavoidable consequences.[68]

In enacting statutes that limit the liability of farmers who violate common-law rules, governments generally shift decision-making authority from the courts to the legislatures. In some cases, legislators exercise their authority directly, through laws and regulations. Other times, they appoint administrative bodies – often operating under explicit political direction – to resolve disputes over agricultural practices. The results are often strikingly different from those obtained under the common law.

Some of the most sweeping examples of governments authorizing nuisances that would be forbidden under common-law rules can be found in provincial right-to-farm laws. First introduced in Manitoba in 1976, right-to-farm laws now exist, in one form or another, in every province. As discussed in the following chapters, such laws have deprived individuals of the common-law property rights that had for centuries proved invaluable in controlling pollution from farms. They have enabled farms to grow beyond the confines maintained by the common law and to generate levels of pollution that the courts would have deemed intolerable.

3 Siding with the Farmer: The Evolution of the Right to Farm in Manitoba

For some it may be an offensive smell, but for some ... it may smell like money.
– Rosann Wowchuk, NDP MLA, 1997[1]

In 1958, Michael and Carolyn Lisoway bought a twenty–acre property in Springfield, Manitoba, ten miles from downtown Winnipeg.[2] Three years later, Leo Clement and Aaga Christensen, doing business as Springfield Hog Ranch Ltd, purchased the adjacent property. The Lisoways fought plans for a large hog farm next door, but failed to prevent the issuance of a building permit. The following year, the farmers built a barn for 1,200 hogs. Five years later, they extended the barn to accommodate 2,000 hogs at a time – 4,000 over the course of a year. The farmers also constructed three large sewage lagoons in which they stored the hogs' waste until spraying it on their crop land.

The Lisoways found the stench of the lagoons intolerable. It drove them inside on even the hottest days, forced them to keep their windows closed, fouled laundry, spoiled butter and other odour-absorbing foods, curtailed entertaining, and made their home a source of embarrassment rather than pride. By 1964, they could take it no longer and tried to sell their home. Not a buyer was to be found.

The Lisoways were not the only people plagued by the odours emanating from the hog farm. The municipal council received frequent complaints about the odours, which spread as far as several miles. Five families successfully appealed their 1968 tax assessments on the basis of the impact of the hog farm. But the stench remained.

In 1971, the Lisoways tried a new approach: they complained about the hog farm to Manitoba's Clean Environment Commission, which had been established three years earlier to supervise environmental preservation and to prevent and control environmental contamination. After a two-day hearing, the commission ordered the farm to reduce the number of hogs from 2,000 to 800 within one year and to eliminate the lagoon system of waste disposal within thirty months. The following year, the Municipal Board dismissed the farm's appeal of the commission's decision.

The government of the day – led by the New Democratic Party's Edward Schreyer – came riding to the farm's rescue. Rather than enforcing the commission's order, it amended the Clean Environment Act to subject commission orders to appeals to the government. The amendments gave the minister 'general supervision and control' over all matters relating to environmental contamination, along with the authority to stay, cancel, vary, or replace orders made by the commission.[3] In 1973, the government went further, passing a regulation exempting all livestock production operations from commission oversight.[4] Those two changes paved the way for an Order in Council freeing the Springfield Hog Ranch from the corrective actions imposed by the commission.[5] Reasoning that any specific limits on odours would be arbitrary, the government decreed that the ranch need no longer reduce its hog population to 800 nor eliminate its waste lagoon. The order effectively shifted the burden of controlling odours from the hog ranch to its neighbours. If the area residents did not like the operation, the government maintained, they could pay for the cost of moving it.[6] This series of regulatory changes marked the beginning of the government's efforts to gain sole decision-making authority over agricultural pollution – to wrest it not just from affected individuals but even from its own environmental agency. The latter charged that these moves could threaten environmental quality, calling the new regulatory regime 'wholly inadequate,' noting that it would not reduce 'malodours' from large livestock operations, and warning that such odours posed a serious and growing problem.[7] Its concerns fell on deaf ears.

The Lisoways fought back: they sued the farm for nuisance. The court's 1975 decision perfectly exemplified the workings of the common law. Justice Wilson examined the facts of the case, considered the positions put forward by the plaintiffs and the defendant, reviewed legal precedents and the principles that had grown out of them, applied the principles to the case at hand, and concluded that the odours from the hog farm did indeed constitute a nuisance. He awarded $10,000 in damages to the Lisoways and ordered the farm to abate the nuisance within eight months.

In considering whether or not the odours constituted a nuisance, Justice Wilson first turned to *Salmond on Torts*, an authoritative legal text that defined nuisances as 'interferences for a substantial length of time by owners or occupiers of property with the use or enjoyment of the neighbouring property.'[8] He then qualified that definition, citing a principle articulated in 1851 by an English court: in order to be a nuisance, something must inconvenience not only sensitive people but also ordinary people. The inconvenience was to be judged 'not merely according to elegant or dainty modes and habits of living but according to plain and sober and simple notions among the English people.'[9]

Justice Wilson further qualified the definition of nuisance, noting that whether something is a nuisance depends on the neighbourhood in which it occurs. Returning to *Salmond on Torts*, he quoted:

> The standard of comfortable living which is thus to be taken as the test of a nuisance is not a single universal standard for all times and places, but a variable standard differing in different localities. The question in every case is not whether the individual plaintiff suffers what he regards as substantial discomfort or inconvenience, but whether the average man who resides in that locality would take the same view of the matter. He who dislikes the noise of traffic must not set up his abode in the heart of a great city. He who loves peace and quiet must not live in a locality devoted to the business of making boilers or steamships.[10]

The justice cited a similar conclusion from an earlier Manitoba case: 'There is no doubt that the nature of the occupancy of a locality may be a large factor in deciding whether the carrying on of certain trades there would or would not create a nuisance.'[11] He extended this principle to the case at hand: 'A corollary, of course, is that one accustomed only to the atmosphere of a downtown or residential metropolitan area cannot be heard overly to complain if, choosing abruptly to move to a rural district in which animal raising is a common practice, the smells are sharply different.'[12]

In his decision, the justice reviewed another principle of the common law: neighbours must make concessions to one another. They must follow 'a rule of give and take, and live and let live,' remembering that 'a balance has to be maintained between the right of the occupier to do what he likes with his own, and the right of his neighbour not to be interfered with.'[13]

The justice then turned his attention to the claims the hog farm had made to try to justify its odours – the first being that it had taken all reasonable steps to avoid producing the smells. 'It is no defence that all possible care and skill was used to prevent the operation complained of from amounting to a nuisance, if nuisance in fact it is. The right to enjoy one's property free from nuisance is an incident of possession independent of the care or want of care of those whose is the nuisance, and does not require proof of negligence.'[14] Even if there were no known way to control a nuisance, a nuisance it would remain. In the words of Justice Wilson, 'plaintiffs and others so affected, not sharing the economic advantages of defendant's operation, are not required meekly to suffer the environmental disadvantages likewise thereto associated, pending resolution of the uncertainties of scientific knowledge.'[15]

Nor, the justice explained, can one who creates a nuisance defend himself on the grounds that his activity falls within zoning regulations or was otherwise authorized by a municipal council. He elaborated, 'Indeed, if in fact what is done gives rise to a nuisance, the defendant may not plead the authority of a statute under which the operation is carried on, unless the nuisance was expressly or impliedly permitted by the legislation, or was the inevitable consequence of that which the statute authorized and contemplated.'[16]

Nor can a nuisance creator defend himself on the grounds that his activity benefits society. As Justice Wilson explained, 'plaintiffs' rights may not be overborne by the social utility of defendant's operation.' He went on to quote a 1930 English decision that had been cited in a famous Ontario case in 1948: 'If I were to consider and give effect to an argument based on the defendant's economic position in the community, or its financial interests, I would in effect be giving to it a veritable power of expropriation of the common law rights of the [neighbouring] owners, without compensation.'[17]

Having reviewed the principles of the common law, Justice Wilson applied them to the facts in the dispute between the Lisoways and the hog farm. He determined the farm's lagoons to be 'a source of odours which, barely acceptable at the best of times, were disastrously offensive in times of high humidity or when the water surface was disturbed by, say, wind effect, or when this fecal matter was drained away, to lie upon the land until so much as had not disappeared by evaporation sank or was ploughed into the soil.'[18] Concluding, 'In the instant case I must endorse plaintiffs' reluctance to accept their state of affairs,' Justice Wilson awarded damages to the Lisoways and issued an injunction prohibiting the hog farm from operating in such as way as to cause a nuisance.

The decision in *Lisoway v. Springfield Hog Ranch Ltd* enraged the provincial government, which resented the censure of an operation it had gone to such lengths to protect and, more generally, feared the potential impacts on the agricultural industry. It lost no time in ensuring that future courts could not issue similar decisions. In May 1976 – just six months after the release of the *Lisoway* decision – it introduced Bill 68, the Nuisance Act, protecting farms and other businesses from common-law liability for the odours they generated. The bill stated that, as long as he does not violate a land-use control law, the Public Health Act, or the Clean Environment Act, a person carrying on a business 'is not liable in nuisance to any person for any odour resulting from the business and shall not be prevented by injunction or other order of a court from carrying on the business because it causes or creates an odour that constitutes a nuisance.'[19]

The debate about Bill 68, one former student recalls, was later used at the University of Manitoba's law school to illustrate the low level of debate in the Legislative Assembly.[20] In introducing the bill for second reading, Attorney-General Howard Pawley spoke of the urban-rural conflicts increasingly caused by Winnipeg's expansion. He called for better land-use controls and planning legislation to minimize future conflicts. But existing farm operations, he said, deserved protection from complaints by their non-farming neighbours. Those who breached municipal, provincial, or federal laws might indeed be held accountable, but not those who violated the common law. As he explained, if an 'action is to be brought I think it should be based upon some clear, precise, defined breach of existing statute or law, not on the basis of the ancient law of nuisance.'[21]

Minister of Mines Sidney Green displayed even greater contempt for the common law. Springfield Hog Ranch, he insisted, 'was not doing anything wrong' and 'was disobeying no laws whatsoever.' Apparently the minister – a Queen's Counsel who presumably had a solid grounding in torts – did not consider the common law worthy of mention. Instead, he determined that since the government had in 1973 reversed the Clean Environment Commission's order against the hog farm, the farm was 'behaving perfectly properly.' Mr Green was incensed that Justice Wilson, who had dared object to the 'ministerial interference' in the matter, had ignored the province's wishes. 'That's why this bill is before the House,' he explained. 'It's before the House because we feel that there has been judicial interference with what has been the decision of the Legislature ... The courts expressed disapproval with what we had done. Therefore we are now expressing disapproval the other way.'[22]

Several members of the opposition commended the government for bringing in the legislation. Progressive Conservative Warner Jorgenson echoed the government's contention that the hog farm 'complied with every law of the land,' apparently forgetting that the common law had governed land use in Manitoba for far longer than current statutes. He called the *Lisoway* decision 'unfair' and concluded that the proposed Nuisance Act was 'needed to prevent further miscarriages ... of justice.'[23]

Throughout the entire legislative debate on the bill, only one member – Liberal Gordon Johnson – expressed reservations about denying people a right to go to court.[24]

The Nuisance Act became law in June 1976. For the first time in Canada, farmers creating unreasonable odours would not be liable in nuisance.[25] Their neighbours could no longer take them to court to obtain damages to compensate them for harms they suffered; nor could they obtain injunctions ensuring that unreasonable odours would cease. The law marked a fundamental shift in Canadians' rights. Rural residents lost their age-old right to enjoy their property, free from disturbing odours. Farmers, on the other hand, gained a new right – one that came to be known as a right to farm. In the following years, every other Canadian province would follow Manitoba's lead.

For sixteen years, Manitoba's farmers sheltered under the Nuisance Act. This shield, however, was insufficient to assure their unfettered operations. In 1992, the provincial government – led by the Progressive Conservatives – extended the shield with the Farm Practices Protection Act. The new legislation, modelled on laws passed by several other provinces, expanded the right to farm, exempting farmers from liability not just for odour but also for noise, dust, smoke, and other disturbances. As long as they used 'normal' farm practices and did not violate laws controlling land use, the environment, or public health, farmers would not be liable in nuisance and would not be subject to court orders enjoining their operations.

The Farm Practices Protection Act also established a government-appointed body – the Farm Practices Protection Board – to resolve disputes between farmers and their neighbours. Those disturbed by farms would have to take their complaints to the board before filing nuisance suits in court. The board would attempt to mediate disputes. If unsuccessful at mediation, it would determine, with the aid of government guidelines, whether the disturbances resulted from normal practices. The board would endorse normal practices and could order the cessation or modification of practices that were not deemed normal. A party could take his complaint to court if dissatisfied with the board's decision. In the event of a subsequent nuisance suit, however, the court would be required to consider the board's decision regarding the matter.

In introducing the new legislation, Minister of Agriculture Glen Findlay noted that as farmers had become the minority in many rural areas, conflicts with their neighbours had escalated. Complaints had led to lawsuits – in some cases targeting normal, 'acceptable' operations that generated unavoidable disturbances. Agriculture, he continued, 'is an important multibillion dollar industry in Manitoba and farmers require protection and assurances that they will be fairly treated ... This proposed legislation is one more effort to support land use in rural Manitoba and ensure continued viability of Manitoba's agriculture industry.'[26]

The New Democrats, now in opposition, were restrained in their support of the bill. Although agreeing with the principles behind the bill, several members expressed concerns about the definition of normal. Rosann Wowchuk wondered, 'What is normal farm practice?' The practice of burning stubble, she pointed out, may be common, but it 'is not the best practice.' Not only does it create smoke that interferes with others, but it also depletes the soil. Similarly, continuous summer fallow can create dust problems for neighbours and cause erosion. 'What are the guidelines,' she asked, 'that this board will follow?'[27] Not surprisingly, she and her colleagues demanded government-made, rather than court-made, answers to such questions, calling for centralized standards, regulations, and guidelines.

Regardless of their insistence on the need for more extensive regulation, NDP members reiterated their support for farmers. 'We are dealing with a way of life here,' Steve Ashton reminded the assembly. 'We are not dealing strictly with a business.'[28] John Plohman made clear his sympathy for farmers facing 'nuisance charges that arise through no fault of their own.' The court process, he said, is 'a long drawn-out process which is not very satisfactory to the farmers, nor fair to those farmers especially ... those who are conducting their farming operations in ... a normal fashion. So it is necessary then to have another system, another way of dealing with this kind of problem.'[29]

The Liberals, too, supported the bill while calling for a clearer definition of normal practices. Not one member echoed the party's earlier concerns about eroding people's rights to go to court.

The Farm Practices Protection Act came into force on 31 January 1994. The intervening years have seen several changes to it. In 1997, the Progressive Conservative government amended the act to give greater force to the decisions handed down by the Farm Practices Protection Board. If a person failed to comply with an order of the board, the board could file a copy of its order in court, giving the order the legal enforceability of a judgment of the court.

The amendment reflected a recognition of the shortcomings of the right-to-farm regime that the government had earlier created. Minister of Agriculture Harry Enns acknowledged that the regime had failed to control the ill effects of farms. As initially established, he explained, the Farm Practices Protection Board could not require farmers to modify unacceptable practices. Although it could order modifications, it could not enforce its orders.[30]

The legislative committee reviewing the amendment heard from a community association in southeastern Winnipeg about the frustrations of relying on a powerless Farm Practices Protection Board. Association president Jim Shapiro described the community's fourteen-year struggle with a farmer who deposited large loads of rotting food and vegetation on the ground for his unpenned pigs – and rats – to eat. The resulting stench made it impossible for his neighbours to sit on their verandas, enjoy their backyards, or entertain guests. Thirty-one people had filed a complaint against the operation in 1994. The board had ruled that the operation did indeed violate normal practices and gave the operator six months to remedy the violation. The farmer refused to comply. As Mr Shapiro explained, 'A glaring deficiency in the Farm Practices Protection Act immediately became apparent. Without an enforcement section to the act there is nothing to compel an individual who is benefiting by not using normal farming practices to change his or her practices to their own disadvantage.'[31] Mr Shapiro complained that an impotent board was not merely useless; it was, in fact, counterproductive, since other municipal or provincial agencies, expecting the board to make and enforce rulings, refused to take action on their own.

Even while proposing to give the Farm Practices Protection Board additional teeth, the agriculture minister stressed his commitment to expanding the livestock industry in the province. Echoing the mantra of successive governments, Mr Enns intoned, 'agriculture is and continues to be of tremendous importance to the economic well-being of this province.' He reminded his colleagues of the loss of the Crow rail subsidy – a loss that had increased transportation costs, reduced the extra-provincial market for grains, and thereby lowered the cost of feed to provincial livestock producers. 'We simply have to find ways and means of utilizing the lower value but high volume feed grains that our farmers are capable of producing in this province, and the obvious choice is to add value to those feed grains through various forms of livestock.'[32]

The NDP's Rosann Wowchuk agreed that increased transportation costs had quite rightly promoted the expansion of the province's livestock industry. 'We are seeing much more livestock being produced in this province,' she told the Legislative Assembly, 'and so we should.' Although she acknowledged that the expanding hog industry in particular had occasioned conflicts over 'the most fine fragrances,' she called for further government intervention in the planning process 'so that we can see the industry grow.'[33] Ms Wowchuk also called for more government research into waste handling and odour control.[34]

The Liberals, too, stressed their support for both the agricultural industry and the right-to-farm regime that supported it. 'Agriculture continues to be important to the economic development of Manitoba,' said MLA Neil Gaudry, adding that 'the Farm Practices Protection Board has played an important part in contributing' to that development.[35]

The tri-partisan support for the amendment reflected a persistent but misguided belief in the importance of agriculture to the provincial economy. Agriculture's alleged economic contribution has run as a common theme through all of the debates on successive right-to-farm acts and amendments. Yet even those making the argument – such as the province's Livestock Stewardship Panel, which has called agriculture 'key to Manitoba's economic future'[36] – rely on data that suggest that agriculture's economic contribution is in fact quite modest. According to that panel, net farm income, excluding government payments to producers and rebates, contributed less than 3 per cent to the provincial GDP in six of the ten years of the 1990s.[37]

More comprehensive data describe an even less productive industry – one that has long been a drain on the economy rather than a contributor to it. Research conducted by Toronto's Urban Renaissance Institute, based on data from Statistics Canada, suggests that between 1992 and 2001, for every dollar earned by a Manitoba farmer, Canadian consumers and taxpayers provided approximately $2.83 in subsidies.[38] Perversely, many of those subsidies have been paid by people far poorer than the farmers they support. In 1997, the average Manitoba farmer had a net worth of $592,600.[39] Adding insult to injury, the artificially low prices resulting from subsidies have tended to benefit consumers outside of the province. Manitobans consume just 11 per cent of the pork produced in the province, with the balance being exported to the rest of Canada and to more than twenty countries, led by the United States and Japan.[40]

Regardless, Manitoba's legislators have never expressed concerns about the industry's failure to pay its way. Indeed, their consistent support for right-to-farm legislation has reflected a willingness to expand the realm of assistance to include environmental subsidies. Nowhere in the legislative debates of 1976, 1992, or 1997 did any party suggest requiring farmers to bear the full environmental costs of their operations. All of the parties were happy to shift those costs to farmers' neighbours and communities.

Four years later, a new series of legislative debates indicated that the parties' support for this unjust and unsustainable approach continued unabated. In 2001, the provincial government – back in the hands of the New Democrats – again amended its right-to-farm law, this time to give the Farm Practices Protection Board the authority to revisit, amend, or revoke orders issued previously. The farming industry supported the changes, despite some reservations about how they might be applied. In a presentation to the legislative committee reviewing the proposed amendments, the Keystone Agricultural Producers, the province's main farm lobby organization, noted that the new rules 'can certainly have a positive effect on the industry, if the request [to replace an order] originates from the agricultural industry.' On the other hand, it worried about requests that might come from outside the industry and warned, 'We know how vexatious some organizations can be in trying to deter the industry.' It proposed a substantial non-refundable application fee to discourage 'unwarranted complaints by uninformed persons without any scientific basis.'[41]

The Manitoba Pork Council, representing 1,500 hog farmers, also made a presentation to the committee reviewing the proposed amendments. The council praised the Farm Practices Protection Act for 'fulfilling its purpose and doing it well.' It praised the 'good judgment and prudence' of the Farm Practices Protection Board, with which it enjoyed a 'strong working relationship.' It noted that, in the seven years the act had been in place, complaints to the board had resulted in only one ceased operation. Like the Keystone Agricultural Producers, the Pork

Council supported amending the act to give the board authority to amend previous orders. And like Keystone, it hoped that changes would be limited to those benefiting farmers. It explained:

> It is our understanding that such amendments would take place to enable the farmer to improve his or her operation by adopting new technology or advanced management practices. We assume the board will use its good judgment in changing, replacing or revoking orders to conform to the intent of the act. A good part of that judgment, hopefully, will factor in the economic viability of effecting a change. We would hope that judgments would also deal accordingly with frivolous or vexatious requests for order amendments. Hog farmers are already being responsible stewards and need not be saddled with additional excessive expenses.[42]

The official opposition seemed especially sensitive to the need to keep farmers' expenses in check. Jack Penner, the Progressive Conservative Critic for Agriculture, noted that Manitoba's farmers had to compete with their generously subsidized counterparts in both the United States and Europe. The Farm Practices Protection Act, he reminded the Legislative Assembly, was initiated 'to ensure that farmers would be able to operate in rural communities without undue intervention by those that have no interest in agriculture.' 'Today,' he continued, 'sometimes one has to wonder how far we have gone in allowing those outside interests to control, to cause regulatory action, to cause even legislative action to occur without giving due consideration to how farmers must, in today's environment, meet the competition.'[43]

Mr Penner urged the agriculture minister to enable farmers to compete, 'whether it is dealing with environmental laws or operational laws or indeed whether it is dealing with financial matters.' Indeed, he suggested, the Farm Practices Protection Board 'at times needs to be given even *more* authority to ensure that the ability of farmers to do their business ... be secured.'[44] Abandoning all pretence of seeking to balance the interests of farmers and their neighbours, Mr Penner asked the minister 'to ensure that the livelihood of the agricultural community comes first, and those who move into the agricultural areas who have no vested interest in the agricultural community become a secondary consideration when we make decisions on livestock operations, when we make decisions on drainage matters.'[45]

Few could have agreed more wholeheartedly than former agriculture minister Harry Enns. The Farm Practices Protection Act, he said, 'really is and needs to be – and it is more important perhaps now than when it was first introduced – a kind of farmer's right-to-farm bill.' The Farm Practices Protection board, he explained, 'is meant to be an advocate for the farmer, for agriculture. There should be no confusion about it. This is not a neutral board that comes to make judgment on a quarrel that is taking place between two neighbours or something like that. If in doubt, this board sides with the farmer.'[46]

Mr Enns' description of a board that is openly biased in favour of agricultural interests contrasts sharply with that provided by the province's department of Agriculture, Food, and Rural Initiatives. The department prefers to present the Farm Practices Protection Act as a mechanism designed to serve farmers and their neighbours alike: 'The Act is intended to provide for a quicker, less expensive and more effective way than lawsuits to resolve complaints about farm practices. It may create an understanding of the nature and circumstances of an agricultural operation, as well as bring about changes to the mutual benefit of all concerned, without the confrontational and unwarranted expense of the courts.'[47] Sadly, this description contains more spin than substance.

Mutual benefit was certainly not prominent in Mr Enns's recollection, in 1997, that part of the rationale for introducing right-to-farm legislation was 'not to have my farmers harassed by well-meaning but insensitive urbanites who ... move out into the country.'[48] Nor does it support his comment, four years later, that people 'are welcome if they want to come and live in the countryside, but not in a trivial way to seriously hinder agriculture.'[49]

Mr Enns's wish has been granted: under Manitoba's right-to-farm regime, rural residents have not been able to seriously hinder agriculture. Livestock operations have grown apace. Between 1975 and 2003, hog production in the province increased more than eightfold, rising from 870,000 to 7.3 million hogs.[50] While the number of hog farms has fallen, the size of the remaining farms has increased dramatically. In 1976, the average farm produced fewer than 200 hogs. By 2001, that number had increased to more than 1,500.[51]

Right-to-farm legislation, although not a direct cause of this intensification, was a necessary condition for it. Many intensive livestock operations have created nuisances that the common law would have forbidden. Were it not for right-to-farm laws, neighbours would doubtless have obtained from the courts injunctions requiring such operations to alter offensive procedures or, in some cases, to scale back operations or shut down entirely.

Neighbours would almost certainly have sought injunctions against the unpleasant – and, it is increasingly believed, unhealthy – air emissions typical of large hog farms. Concerns about odours from farms were common even before livestock operations became so concentrated. In the late 1970s, approximately 75 per cent of the complaints received by provincial environmental authorities concerned odours from livestock operations.[52] The sheer volume of manure that large farms produce makes conflict with neighbours more likely. A large hog produces more than 2000 kilograms of manure a year, and the smell of hog manure is notoriously disagreeable. A farm with several thousand hogs – especially one that manages its manure using so-called normal methods – is likely to be a smelly place indeed.

Successive governments have operated under the belief that even the best-intentioned farmers could not eliminate emissions from their operations.[53] That belief has been something of a self-fulfilling prophecy, since the right-to-farm laws growing out of it have curbed incentives to develop innovative solutions. In the absence of such laws, the demand for effective technologies would surely have spurred innovations in manure management and treatment. As the province's own Livestock Stewardship Panel pointed out, 'Market forces drive the improvement of breeds, the improvement of feed and weight gain ... and similar effects in a fairly direct manner. It can be expected that a tight, interactive link will be maintained between operators and researchers, backed up by industry organization without much intervention by a third party such as government ... When one is focussed on the animal and the bottom line, the industry needs little prodding to support and adopt new techniques.'[54] The panel noted that the market does not tend to drive environmental stewardship in the same way. And how could it? Right-to-farm laws have taken many environmental decisions out of the marketplace. They have taken negotiation and decision-making authority out of the hands of affected individuals and placed them in the hands of politicians and bureaucrats. The environmental and social costs of such disempowering laws have been significant.

4 Raising a Stink: The Legacy of Right-to-Farm Legislation in New Brunswick

We don't want to farm by law.
    – New Brunswick Pork Producers Association, 1985[1]

Metz Farms 2 Ltd, an enormous piggery that operated for almost six years just outside Sainte-Marie-de-Kent near New Brunswick's eastern coast, was among Canada's most controversial farms. It housed 10,000 hogs at one time – 30,000 over the course of a year – and produced 24 million litres of liquid manure in a year. Whether stored in the farm's 90–by-90 metre lagoon or sprayed on nearby fields, the manure smelled. Neighbours described the odours from the farm as 'unbearable' and 'totally nauseating.' Some complained of tearing eyes, sore throats, and weeping sores, others of being kept from their gardens. In the words of one local resident, 'We've never experienced anything that smells so bad. People shouldn't be forced to accept the introduction of something like this in their community. It's hell.'[2]

In nearby Bouctouche – voted one of Canada's ten prettiest towns by *Harrowsmith* magazine – the mayor worried that the smells from Metz 2 drove tourists away. The chief of the Bouctouche First Nation likewise complained that tourists visiting his band's cultural centre found the farm's smells offensive.

Area residents were concerned not just about the farm's stench but also about the air and water pollution often associated with manure. The farm's manure, they feared, would contaminate the air with ammonia, hydrogen sulfide, and methane, and contaminate the water with pathogens that could threaten both human health and the sensitive oyster fishery in Bouctouche Bay. Local alarm increased when a microbiologist found that, after rainfalls, fecal coliform levels soared in waters near fields that had been sprayed with Metz 2's manure.[3]

Concerned citizens worked relentlessly to persuade provincial politicians to shut down the operation. Indeed, they began their protests even before the farm commenced operations in 1999, first requesting an environmental impact assessment, and later applying (unsuccessfully) for a judicial review of the government's refusal to require one. In the following years, citizens organized motorcades and rallies, occupied government offices, gathered thousands of names on petitions, and went to court to demand the release of information about the farm's impacts on the environment. One organization, the Association for the Preservation of the Bouctouche Watershed, spent $250,000, raised from 3,000 people, opposing the farm.

Some of Metz 2's opponents also put pressure on the farm itself, at times resorting to illegal tactics. Efforts to block a tanker truck of manure resulted in a court injunction against the protesters. But the injunction didn't bring social peace: Metz 2's manager alleged that his employees and their families were harassed and threatened, that one employee's house was burned, and that family pets were poisoned.

Neither the social discord nor the tireless political pressure undermined provincial support for the pig farm. Indeed, the provincial government – having worked to recruit European farmers who, restrained by tough regulations, could not expand their operations in their native lands – consistently defended the farm.[4] It insisted that it had found no evidence of ground or surface water pollution, and that Metz 2 operated within the bounds of the law. Its defence of the farm was hardly reassuring. A provincial report on water quality in 2000 and 2001 found increased concentrations of phosphorus, nitrogen, nitrate/nitrite, potassium, copper, fecal coliforms, total coliforms, and E. coli in waters near fields where the farm's manure had been applied. It also, however, found elevated levels in areas where the farm's manure had not been applied. It therefore concluded, 'There is no clear evidence that the Metz program *alone* is having a measurable adverse effect on water quality in the study area.'[5]

Agriculture minister Rodney Weston insisted that Metz 2 'has respected every provincial act and regulation they are subject to. They have not broken any rules or regulations.'[6] This would be more comforting if the province regulated odours, the impact most vigorously opposed by the local community. But the province neither monitors nor controls odours. It did commission a report on air quality: the consultant found high odour levels as far as nine kilometres away from Metz 2 and reported that odour levels from the farm greatly exceeded those permitted in other jurisdictions.[7]

Regardless of Metz 2's environmental impacts, both the government and the opposition made it clear that they would rather shore up the farm with subsidies than shut it down. In March 2002, the province announced that it would provide financial assistance to help the farm increase its economic yield and introduce odour-control systems – a gift valued in the press at $1.5 million. Mr Weston called the subsidy 'the only reasonable and realistic approach to maintain[ing] a viable hog industry in New Brunswick while addressing the citizens' concerns.'[8]

The government's commitment to propping up a polluting industry infuriated local residents. Their criticisms did not faze Mr Weston. Environmentalists 'say we're too friendly with agriculture? That's nice to hear that we are friendly with agriculture because we are; we believe there is a strong future for the industry in New Brunswick.'[9] Metz 2 returned the friendly sentiments. Expressing his delight with the subsidy, the farm's manager gushed, 'I couldn't be more pleased with our new minister.'[10]

Unfortunately, the promise of provincial assistance did not solve Metz 2's odour problems. The first steps taken by the farm included the installation of ozonation equipment to purify the air in the hog barn and the smothering of the manure lagoon with a blanket of barley, wheat, straw, and canola oil. While management claimed significant reductions in odours, local residents continued to complain. One told the CBC, 'I can still smell it. In fact, last night, if I couldn't get out of it, I would have been sick on the road.'[11] Long-term plans to reduce odours from Metz 2 by treating its manure failed to materialize. The government remained committed to funding a treatment facility. It consulted with the farm and provided the requisite environmental approvals, but awaited Metz's decision on proceeding.[12] Although the government expected to see a manure treatment facility in place by May 2006, it refrained from specifying plans for such a facility in the farm's operating licence when renewing it in 2004.[13]

Plans for a manure treatment facility – and indeed, for the farm itself – appear to have been scotched by Metz itself. In the summer of 2005, the pigs disappeared from the property. Local residents conjectured that a collapsed roof and low pork prices contributed to the decision to remove the livestock. The agriculture ministry insisted that regulatory pressure played no role – the farm's licence, it noted, remained in effect.[14] Metz 2's manager was tight-lipped about the meaning of the removal, calling talk of the farm's closure '95 percent speculation.'[15] With no reason to believe that the farm had closed permanently, neighbours remained concerned about both the manure stored in the lagoon and the farm's future.[16]

Indeed, the removal of the pigs did not put an end to a lawsuit filed in December 2002 by forty-five local residents. The suit targeted not only Metz 2 but also thirteen other farmers who had spread Metz 2's manure on their farms. The statement of claim alleged contamination of groundwater, surface water, and air, violations of provincial laws protecting public health and the environment, and trespass and nuisance. The plaintiffs sought compensation and an injunction. Local resident Neil Gardner explained, 'Unfortunately, the only person a politician has to listen to is a judge. What we are looking for is justice, and it seems to be the only place where we stand a chance at finding justice is in the courts.'[17]

Mr Gardner's search for justice in the courts sounds far more quixotic today than it would have sounded two decades ago. Until 1986, Metz 2's neighbours could have sued the farm for creating nuisance odours. The courts – rarely tolerant of extreme odours – would have determined whether Metz 2 violated its neighbours' common-law property rights. But this legal option was not available to Metz 2's neighbours. In 1986, following two successful nuisance suits against farmers who created foul odours, New Brunswick's government outlawed such suits. To protect the agricultural industry, the government sacrificed its citizens' age-old rights to use and enjoy their property.

The story begins in Charlo, New Brunswick, with a pig farmer named Terry Sullivan. When Mr Sullivan expanded his farm in 1980, his neighbours complained. His 140 sows and 1,000 feeder hogs produced 260 cubic feet of liquid manure each day, which, channelled into a manure lagoon for storage and decomposition, created an unbearable stench. The community, Mr Justice Ronald Stevenson later said, was 'in a state of siege by smell.'[18] Children could not play outdoors; families could not barbecue, air their laundry, or even leave their windows open. Some complained of nausea, others of sleeplessness and stress. Property values plummeted. The community battled the pig farm on many fronts. Residents complained to the village council and to three provincial departments, with no avail. They even organized a protest parade. Finally, twenty-five neighbours went to court, seeking damages and an injunction.

Justice Stevenson, who heard the case at the Trial Division, found that the pig farm's offensive odours constituted an intolerable nuisance: they interfered with the plaintiffs' reasonable enjoyment of their properties. In a November 1985 decision, he awarded damages, totalling $30,500, to twenty-two of the farm's neighbours. In response to Mr Sullivan's argument that excessive damages could ruin not only him but also other farmers in similar situations, the judge insisted, 'economic considerations are not a shield for the commission of tortious acts with impunity.'[19] Justice Stevenson did not grant an injunction, explaining that he could not enjoin something that did not exist. Since a 1984 fire had destroyed Mr Sullivan's barn and 1214 animals, and since the manure lagoon had been pumped out and covered a year later, the nuisance had ceased. The judge noted, however, that had the nuisance continued, he would have had no alternative but to issue an injunction. And if the operations were to resume (as was threatened by the presence of a new barn and lagoon on the property) they could face an injunction in a new lawsuit. Thirteen months later, the Court of Appeal upheld Justice Stevenson's decision. The Supreme Court of Canada refused to hear a further appeal.

In his decision, Justice Stevenson considered the character of the neighbourhood. No building, zoning, or other land-use by-laws governed the area. Although the area had at one time been strictly rural, with farming as its principle activity, it had in the last dozen years become a bedroom community for people working in Dalhousie. It had acquired a definite suburban character by the time Mr Sullivan began expanding his operations.

Justice Stevenson also considered the regulatory environment. In his defence, Mr Sullivan had argued that he had complied with the requirements of the provincial departments of health, agriculture, and environment. The judge rejected the argument. He called one regulation's failure to distinguish between small and large operations 'strikingly unrealistic.' In any event, he explained, 'compliance with rules and regulations contained in or authorized by legislation does not relieve one from the common law rule that he must conduct his activities in such a way as not to create a nuisance.'[20] The Court of Appeal agreed. In a decision released in December 1986, it explained that provincial certificates of compliance 'demonstrate Mr. Sullivan's good intentions in attempting to construct a piggery in accordance with approved practices. The evidence discloses that Mr. Sullivan sought and usually followed the advice of officials in the Departments of Agriculture and Environment.

But, of course, he could not, *at least in 1980*, obtain a licence to create a nuisance.'[21]

The Court of Appeal's cryptic reference to 1980 suggested that the law had changed in the intervening years. And in fact it had. After Justice Stevenson issued his decision in November 1985, the farming industry demanded protection against similar suits in the future. Richard Hatfield's Progressive Conservatives were only too happy to comply. It took them less than three weeks to respond to the trial judge's decision: in early December, Minister of Agriculture Hazen Myers promised to fast-track right-to-farm legislation – legislation that his government had already begun contemplating in response to an earlier court case pitting a polluting piggery near Lac Unique against more than seventy cottage owners.[22]

Indeed, a discussion paper drafted by the Department of Agriculture and Rural Development had highlighted that court decision, which had cost the piggery $18,000 in damages – and had cost many other farmers their confidence in the security of their investments. The paper referred to the court's refusal to consider its decision's financial impacts on the piggery's owners, quoting its conclusion that 'the right to a healthy and clean environment cannot give way to pecuniary interest ... Moreover, the time to consider this aspect of the question was before the construction of this piggery when it was known that this lake was a resort where there were already about seventy-five cottages.' The agriculture department's discussion paper, noting farmers' concerns about the principle applied in the Lac Unique case, proposed a 'better balance among the rights of all concerned.' Yet the solution it proposed could hardly be described as balanced: 'the balance should be weighted towards a farmer in rural agricultural areas.' Anticipating an increase in disputes as farms grew and non-farming residents moved to rural areas, the paper called the court decision 'reason to consider whether or not the social and economic conditions have changed sufficiently to warrant a change to this common law by the introduction of a 'Right-To-Farm' statute.' The foundation of its argument? Noises, dust, smells, and traffic were 'a necessary part of the operation of the agricultural industry. Without them there would be no food in our supermarkets.'[23]

If the Lac Unique case had prompted quiet speculation about new rules governing agricultural nuisances, the decision against Terry Sullivan imparted life to the issue, giving it unprecedented popular and political urgency. The matter quickly went to the Select Committee on Agriculture and Renewable Resources, where Deputy Agriculture Minister Tim Andrew claimed somewhat melodramatically that the Sullivan case was 'very worrying to anybody who expects to continue to eat.' The decision, he insisted, could put every farmer at risk and cause immeasurable damage to the province. He continued, 'The only way we can protect the farmers of New Brunswick from this kind of problem is to put into place a legislative framework as an alternative to the common law.' Mr Andrew urged the committee to recommend right-to-farm legislation based on Manitoba's 1976 Nuisance Act, perhaps expanding the permitted nuisances to include not only odours but also noise and dust. He called his proposal 'right-to-eat legislation,' warning, 'If we don't have food, we are liable to go hungry.'[24]

The select committee heard presentations from dozens of agricultural organizations, all demanding special protections. The Sullivan case was very much on the minds of the presenters. The New Brunswick Federation of Agriculture expressed its outrage over the case, telling the committee, 'This man's livelihood is being held up for ransom.' Recent court cases, it complained, set 'a pattern which says clearly that farmers do not have enough protection under the law.' The federation called for 'broad protection' – broader, even, than that offered in Manitoba.[25]

The New Brunswick Potato Agency complained that 'recent developments have shown that the present judicial system is unjust against the farming community.'[26] The New Brunswick Egg Marketing Board warned that no farmer would be safe from lawsuits in the future: 'You get two or three of these lawsuits going and everybody jumps on the band wagon ... Whether you are a hog man, dairy man, or poultry man, whatever you are, you cannot laugh and say, "that is not me," because tomorrow it may be.'[27] In expressing its own concerns about the Sullivan case, the New Brunswick Chicken Marketing Board sounded a common theme – that farmers were being unfairly singled out while other industries polluted with impunity: 'Everybody has the right to make all the odours they want in the province, they pollute their rivers or streams and do whatever they want, but we are the ones who are not allowed to do anything.'[28]

Presenters raised a variety of other issues as well. The New Brunswick Milk Marketing Board stressed the need to maintain farmland. It played the jobs card, citing potential losses not only in farming jobs but also in those associated with processing and transportation. And it suggested that non-farm residents should absorb the costs associated with living in rural regions. Referring to the work of economist A.C. Fisher, it told the committee that farmers' neighbours should not be compensated for nuisances. Rather, 'it makes more sense to have the nonfarm rural resident compensate or pay the farmer if he wants a reduction in odour, dust, and noise, because he created an increase in the damage done by moving close to the farm operation.'[29]

The New Brunswick Soil and Crop Improvement Association likewise argued that farmers should not bear the full costs of improved manure management. It advocated general taxpayer support, in the form of low-interest loans, on the grounds that the changes would benefit all of society. It also insisted that farmers could not afford the time and money to fight court battles. In its words, 'We are farmers, not lawyers, and we need "Right-to-Farm" legislation to guarantee us that we can go on doing what we do best.'[30]

And what if farmers were to remain subject to legal actions? The bleakest prophecy came from the Restigouche Farm Women's Association: 'Without a viable agricultural industry, life in this country would be non-existent.'[31]

One of the most shameless pleas came from the New Brunswick Pork Producers Association, which asked the committee for extraordinarily broad exemptions from legal controls of any kind. It requested protection not only from the common law but also from restrictive statutes and regulations, explaining, 'we don't want to farm by law.'[32]

Other presenters described an industry already free of meaningful restrictions. District Medical Health Officer Médard Bérubé complained that the government refused to enforce provisions in the Health Act requiring his authorization of piggeries prior to their construction. He decried his illusory ability to prevent the construction of a piggery that threatened public health: 'Not only was my authorization not asked for, the project continued despite my best efforts to block it ... We appear to have more power than we actually do.' Furthermore, he criticized the province's out-of-date laws and regulations, explaining that provisions designed to regulate small family farms were ineffective in controlling operations with a thousand or more animals.[33]

The committee also heard from non-farming rural residents opposed to right-to-farm legislation. A number of citizens reiterated their case in the provincial press. Several noted that they had lived in their homes long before new farms intruded, and that it was they who needed protection from farms rather than vice versa. They bemoaned the futility of relying on unenforceable manure management guidelines to prevent pollution from farms. They described their frustrating efforts to contact the departments of health, agriculture, and environment, visit the provincial ombudsman, petition their MLA, and even involve the RCMP. All ignored their pleas. As one rural resident complained, 'We can say anything we want and talk until we're blue in the face, but no one will listen to us or help us.' His wife added, 'It seems the pigs' votes are stronger than ours.'[34] The implication was clear: rural residents themselves were the only ones likely to hold farmers accountable for their operations. And it was their common-law rights that enabled them to do so.

These arguments for empowering rather than disempowering rural residents fell on deaf ears. In May 1986, the agriculture minister introduced Bill 64, the Agricultural Operation Practices Act (AOPA), to protect farmers from common-law actions. Under the act, 'A person who carries on an agricultural operation and who ... does not violate any land use control law, the Health Act ... or the Clean Environment Act ... is not liable in nuisance to any person for any odour, noise or dust resulting from the agricultural operation and shall not be prevented by injunction or other order of a court from carrying on the agricultural operation because it causes or creates an odour, a noise or dust that constitutes a nuisance.' The act also specified that the onus of proving that a farmer violated one of the above-mentioned laws would lie on the plaintiff.

In one important regard, the AOPA exceeded the demands of many who had appeared before the Select Committee on Agriculture and Renewable Resources the previous winter. The law did not distinguish between existing and new farms, nor did it distinguish between farms surrounded by farmland and farms surrounded by residences. Many individuals and organizations – both those arguing for and those opposing right-to-farm legislation – had advocated a system of 'first in time, first in right.' The chairman of the select committee, reviewing farmers' widespread fears of being sued by new neighbours for long-established practices, acknowledged that a consensus existed among presenters that 'if the shoe is on the other foot, they also agree that private citizens should be protected from farmers who move to residential areas.'[35] Despite this consensus, the law as enacted offered no protection from agricultural nuisances for people living in residential areas.

The government presented the AOPA as one 'designed to provide security to farmers.'[36] The opposition approved of both the end and the means. Liberal MLA Douglas Young called the legislation 'very important' and 'appropriate,' maintaining that 'we have a responsibility to the agricultural community to give them some degree of certainty as to how they can function and assure them that they will not be subject to problems that could cause them a lot of difficulty.' 'Every New Brunswicker will agree,' he suggested, 'that we had to have some type of legislation that would provide for a degree of long-term stability for farmers.' Although odours from farms 'can make life very uncomfortable,' he acknowledged, 'there has to be an accommodation.'[37] New Brunswick's New Democratic Party also favoured protecting farmers from their neighbours. In fact, two weeks after the announcement of the court decision against Terry Sullivan, the party had passed an emergency resolution upholding the right of farmers to apply manure to their lands.[38]

Unopposed, the AOPA was assented to on 18 June 1986, and was proclaimed and came into force the following day. From that day on, New Brunswick's courts could no longer apply the ancient maxim, 'use your own property so as not to harm another's,' to farmers whose odours, dust, or noise harmed their neighbours.

The farming industry did not for long remain satisfied with its freedom to create offensive odours, dust, and noise. Many farmers, concerned that the smallest violations of the Clean Environment Act could leave them vulnerable to nuisance lawsuits, pressed for legislation that did not link liability limitations to conformity with statutes.[39] They redoubled their efforts after the bankrupt Terry Sullivan killed himself in 1990. Led by the New Brunswick Federation of Agriculture (since replaced by the Agricultural Producers Association of New Brunswick), they pushed for new legislation providing broader protections.

The Liberal government agreed to a new AOPA, which it passed in March 1999. The new act extended the scope of permitted farming activities. It shielded farmers from liability for an endless variety of nuisances: 'A person who carries on an agricultural operation using acceptable farm practices is not liable in nuisance to any person for any odour, noise, dust, vibration, light, smoke or other disturbance resulting from the agricultural operation and shall not be prevented by injunction or other order of a court from carrying on the agricultural operation.'

In introducing the concept of 'acceptable' practices, the act redefined the factors guaranteeing farmers protection from nuisance suits. No longer would obedience to land-use, health, and environmental legislation protect a farmer against liability. Instead, according to the act, the protection would result from the use of practices that were 'consistent with proper and accepted customs and standards as established and followed by similar agricultural operations under similar circumstances.' Acceptable practices were not just those that other farmers were using; they also included the use of innovative technology and advanced management practices. The act specified that it did not exempt farmers from compliance with any provincial or federal statute or regulation. These provisions, however, appeared in the paragraph *following* that governing nuisance liability. The act did not say that non-compliance would provide grounds for a nuisance suit.

The new AOPA established a Farm Practices Review Board to study farm practices and to resolve disputes between farmers and their neighbours. The board, to be appointed by the Lieutenant Governor in Council, was structured to be biased towards the agricultural community, with four members recommended by farm organizations and just two members who were not farmers. The board would hear complaints about disturbances from farms and determine whether the disturbances resulted from acceptable practices. Ninety days after complaining to the board, an applicant could proceed with a nuisance action. In such an event, however, the act required the court to consider the determination of the board.

The Liberal government, swept from power less than three months after passing the new AOPA, did not proclaim it. That job fell to the Progressive Conservatives, who acted only after several years had passed – and only after the steady publicity over the Metz 2 farm in Sainte-Marie-de-Kent made disputes between farmers and their neighbours increasingly difficult to ignore. In March 2002, agriculture minister Rodney Weston announced his intention to proclaim the new AOPA, vowing, 'we are committed to ensuring the future of the farming community in our province and will do everything necessary to keep that commitment.' He assured the agricultural community of his government's proud support and its desire to promote growth in the industry. The minister justified his support on economic grounds. Agriculture, he claimed, 'has always been a major economic contributor to New Brunswick and contributes greatly to its economic health ... Agriculture plays a vital role in the prosperity of rural New Brunswick. Agriculture supports the employment of many workers in the transportation, manufacturing, retail and service sectors.'[40]

The irony of announcing, in the same release, enhanced financial programs for farmers seemed to be lost on the minister. Also apparently lost on him was the grim economic reality that New Brunswick's inefficient agricultural sector is a money-losing drain on society. Between 1992 and 2001, for every dollar earned by a New Brunswick farmer,

Canadian consumers and taxpayers provided approximately $4.50 in subsidies.[41] The new AOPA would constitute yet another subsidy – unmeasured, but enormous – to the agricultural industry.

Sensitive, perhaps, to the political costs of eroding citizens' cherished property rights, the government repeatedly insisted that proclaiming the new AOPA would not take away any public right of recourse to the courts.[42] This claim was belied by Mr Weston, whose explanation was carried by the CBC: 'If citizen B decides to go to the courts and wants to spend their money and take that farmer to court, they still have that recourse. I'm saying here today they don't have a leg to stand on.'[43] The *New Brunswick Telegraph-Journal* captured the essence of the government's policy with the headline, 'N.B. moves to protect farmers from nuisance neighbours.'[44] In a bizarre perversion of traditional legal, economic, and environmental principles, rural residents, rather than polluting farmers, were deemed responsible for conflicts. And it was they who would be expected to bear the costs of agricultural nuisances.

Since the new AOPA came into force on 15 January 2003, New Brunswick's rural residents have less power than ever to protect their homes, families, and environments. No longer are disputes with their farming neighbours matters for discussion, negotiation, and, ultimately, resolution through the courts. Instead, they are matters to be decided by politicians, bureaucrats, and administrative boards disproportionately representing the agricultural industry. No longer able to turn to courts guided by age-old principles protecting the use and enjoyment of their property, rural residents must now plead their cases in front of those guided by self-interest and standards based on what is normal rather than what is right. The bias towards farmers inherent in the AOPA and the board it established may help explain why the board received just one application during its first three years in existence.[45] New Brunswickers may well have determined that their best hopes for environmental protection now lie not in the law but in political protest. To control a stink, they must now raise a stink.

The manager of Metz 2, who had earlier demanded greater protection from nuisance lawsuits,[46] called the new AOPA 'great news for the agriculture industry.'[47] In contrast, environmental activist and columnist Janice Harvey called it 'a nightmare waiting to unfold.'[48] Both assessments may well prove correct.

5 A Mushrooming Problem: Agricultural Nuisances in Ontario

I stood outside this morning, it was pouring rain and I just drank in the smells of this beautiful spring morning. How precious smells can be. The smells of the earth, the rain, were indescribably wonderful. How wrong it is that [Greenwood Mushroom Farms] could just move in and take all this away from us.

– Mushroom farm neighbour Patricia Pyke, 1996[1]

In July 2004, 150 residents of Ashburn, in Ontario's Durham Region, filed a civil lawsuit against Greenwood Mushroom Farms (GMF), claiming that the stench from the farm created a nuisance. The same day, one resident also launched a private prosecution under the Ontario Provincial Offences Act, alleging that noxious odours from GMF violated the Environmental Protection Act. As well, the issue was in front of the Normal Farm Practices Protection Board – a body established to resolve disputes over agricultural operations.

The presentation of the same complaint to three different forums provides a glimpse into the tangle of laws and regulations that govern pollution from farms in Ontario. This web of ambiguous legislation, conflicting policies, competing jurisdictions, and parallel processes has sown confusion and frustration, leaving farmers and their neighbours alike uncertain of their rights and responsibilities.

Since GMF's establishment in 1994, it has plagued its neighbours with foul odours and a dense haze. The emissions have sickened some nearby residents, burning their eyes, noses, and throats and making breathing difficult. The odours have disrupted their lives in countless other ways, as well. No longer can those living near the farm plan on eating outside, hanging clothes on their lines, or working in their gardens. When the odours are strong, they must close their windows, even in the hottest weather. Their dogs pick up the odours and bring them inside. Their clothes pick up the odours and take them to work. Colleagues comment. Friends and relatives stay away.

The problem does not lie with mushroom farming per se. Rather, it lies with the production of the composted substrate in which mushrooms are grown. At GMF, substrate production begins with the mixing of vast quantities – up to 390,000 pounds a week – of poultry litter, manure-sodden hay from horse stables, corn cobs, straw, and other materials. The mixture is piled on an outdoor slab measuring 620 feet by 92 feet, where it is periodically turned and wetted – and from where it begins to emit smells that permeate the neighbourhood. GMF's owner has taken measures to limit the odours, including moving the mixture into bunkers, where air is vacuumed from the compost and passed through an ozone injector tank before being released. Nonetheless, neighbours complain that the odours persist.

Local residents, many of whom have lived on and farmed their land for decades, have nothing against the odours associated with traditional farming operations. Indeed, few objected to even the pig farm that operated in the area for years. They understood that rural residents occasionally have to put up with unpleasant odours. But these are no ordinary odours. Neighbours liken them to the smells of ammonia, rotten eggs, decaying cabbage, and faecal material. They describe them as noxious, rancid, foul, and horrendous. One former neighbour equated living with the smells to being 'held upside down one inch away from excrement in an outhouse used by 400 people.'[2]

Such revulsion is common among those living near mushroom farms. Composting for mushroom farms is a notoriously unpleasant business. As mushroom expert Danny Rinker has written, substrate preparation 'is becoming a significant issue world-wide.' Offensive odours have led some mushroom farmers to purchase their compost from remote suppliers or to locate their composting operations far from their farms.[3] 'Anyone considering constructing or purchasing a mushroom farm,' Mr. Rinker advises, 'must be especially cautious in choosing the site for compost preparation.'[4]

GMF's neighbours would like the farm to find another, more appropriate, site for its substrate preparation. Alternatively, they would like the farm to fully enclose its operations and to effectively treat all odours. Whatever the means, their ends are clear: in the statement of claim for their civil suit, they requested an injunction ordering GMF to not discharge noxious emissions. They also sought more than $10 million in damages.[5]

The residents of Ashburn have a classic nuisance claim: GMF, they charge, has violated their common-law property rights by interfering with the use and enjoyment of their properties. Until 1988, their recourse would have been straightforward. If they could have convinced a court of the nuisance, they would have been entitled to damages and an injunction – an order requiring the cessation of the offensive activity or specifying corrective action. But things are no longer so simple: now the court, in determining whether the farm is creating an actionable nuisance, must decide whether or not its activities are 'normal.'

In 1988, the Ontario government, under Liberal Premier David Peterson, severely limited farmers' liability for the nuisances they created. It proclaimed the Farm Practices Protection Act, which specified that as long as a farmer did not violate one of several statutes, he would not be liable in nuisance for any odour, noise, or dust resulting from a normal farm practice. Under the act, those disturbed by farming operations could take their complaints to the Farm Practices Protection Board, which would determine the normalcy of a disputed practice. The board would dismiss complaints resulting from normal practices and order changes to those practices that were not normal. The board could not issue damages to those harmed by abnormal practices; that remained the responsibility of the courts.

Although it was widely assumed that farmers required such legislation to protect them from court actions, that assumption rested on very little evidence. The advisory committee set up by the provincial government to consider right-to-farm legislation presented the case of New Brunswick's Terry Sullivan as 'evidence of the lack of support in the courts for farmers.'[6] However, it provided no information regarding the number of farmers that had been sued in Ontario or the kinds of practices that had been challenged. Indeed, it admitted that it had 'had difficulty in getting information on the nature and extent of nuisance complaints.'[7]

The advisory committee's own numbers on disputes between farmers and their neighbours suggest that its concerns may have been overblown. It noted that the Farm Pollution Advisory Committee, which worked to resolve farmer-to-farmer complaints and reported to the environment ministry, had mediated about 100 nuisance disputes in the previous 13 years.[8] Seven or eight cases a year hardly signified serious trouble warranting draconian legislation. Nor did the results of a survey conducted for the advisory committee justify far-reaching legislative protections. The survey, which elicited 222 replies, asked farmers about the kinds of complaints they had received. The most common complaints concerned the movement of machinery on highways. Seventy-six farmers mentioned receiving complaints about odours, manure storage, or manure spreading. They gave no indication, however, that they had been unable to resolve such complaints.[9]

It seems that right-to-farm advocates were concerned more about future threats than past lawsuits. They foresaw increasing land-use conflicts as farmers severed portions of their lands or sold entire properties and as urban folk moved into rural areas. They understood that the increasing size of the farms that remained would likely exacerbate disputes, not just with newcomers but with long-established neighbours. They feared a loss of prime farmland and, with it, a loss of their dream of provincial self-sufficiency in food. They feared threats to our heritage – and even to our national well-being. And they feared for the jobs of the quarter of the provincial workforce that was tied to agriculture.[10]

The Farm Practices Protection Act of 1988 failed to assuage such fears. A decade later, under pressure from a number of agricultural groups,[11] the Progressive Conservative government of Mike Harris replaced the legislation with even stronger fare. Under the new Farming and Food Production Protection Act (FFPPA), the government expanded the list of protected disturbances to include flies, light, smoke, and vibrations. Disregarding the concerns of the environment ministry, it also broke the link between compliance with statutes and protection from civil liability. It did so quietly – perhaps, as one judge later suggested, to avoid offending those who promote environmental protection.[12] In any case, as of 1998, 'a farmer is not liable in nuisance to any person for a disturbance resulting from an agricultural operation carried on as a normal farm practice.' Nor, under the act, need a farmer comply with municipal by-laws that restrict normal farm practices.

And what, exactly, is a normal farm practice? The FFPPA defines a normal practice as one that 'is conducted in a manner consistent with proper and acceptable customs and standards as established and followed by similar agricultural operations under similar circumstances' or, alternatively, as one that 'makes use of innovative technology in a manner consistent with proper advanced farm management practices.' The agriculture ministry offers this test of normalcy: 'Would a farmer with average, to above average, management skills use this same practice on his/her farm under the same circumstances?'[13]

The court hearing the civil case against Greenwood Mushroom Farms will have to determine whether any nuisances it creates result from normal farm practices. The question has been debated in the courts before. In 1995, nineteen neighbours sued GMF for disrupting their lives and the use of their properties. In 1999, trial judge Donald Ferguson determined that odours from the farm's composting process did indeed constitute a nuisance. From the commencement of GMF's operations in 1994, he found, the odours had significantly affected its neighbours' physical well-being and substantially disrupted their use of their lands. The judge called the interference severe, unprecedented for the area, and intolerable. He acknowledged that the mushroom farmers had taken 'all the reasonable precautions possible in the current state of the art of mushroom composting.' 'However,' he concluded, 'the use of the defendants' land for composting is unreasonable having regard to the fact that they have neighbours.'[14]

Justice Ferguson had no trouble deciding that the practice was not normal, both because GMF had initially carried out its composting improperly and because it operated 'in an area where the nuisance it produced was completely out of character.' The intensity and frequency of the nuisance had 'fundamentally changed the rural environment the plaintiffs had enjoyed before.' Thus, he concluded, 'the GMF operation was never in the category of normal farm practice.'[15]

Justice Ferguson awarded $263,500, plus interest, in damages. He would have liked to issue an injunction to stop the nuisance. Yet, the wording of the right-to-farm law is sufficiently ambiguous that he believed – wrongly, it turns out – that he lacked the power to do so. When, two years later, the Ontario Court of Appeal upheld Justice Ferguson's decision, it said that a court does indeed have the power to enjoin a nuisance that has not resulted from a normal farming practice. Since the plaintiffs had not cross-appealed the point, however, the Court of Appeal did not interfere with the remedy ordered by Justice Ferguson.

Usually, it is not the courts that decide whether a farming practice is normal, and therefore acceptable. That determination generally falls to the Normal Farm Practices Protection Board (NFPPB). The board is appointed by the Minister of Agriculture and its decisions must be consistent with any directives, guidelines, or policy statements issued by the Minister. The board holds hearings into complaints. If it finds that the disturbance complained of results from a normal farm practice, it dismisses the application. A party may appeal the decision to the Divisional Court.

During the first trial of GMF, Justice Ferguson, aware of the 'very significant policy-making role in determining what is a normal farm practice,' considered leaving the determination of what is 'normal' to the NFPPB. He was concerned, however, that the parties to the lawsuit had already gone to enormous expense to present relevant evidence to him. He also pointed out that, even if the board were to decide that GMF's practices were not normal, it wouldn't have the authority to award damages to the plaintiffs. If they wanted compensation for their years of suffering, the plaintiffs would have to return to court to obtain damages. For these reasons, Justice Ferguson chose to decide the issue himself.

When GMF appealed Justice Ferguson's decision, the Ontario Federation of Agriculture intervened in the lawsuit, submitting that the NFPPB, and not the courts, should determine whether a farming practice is normal. Only after the board has determined that a practice is not normal, it maintained, should the courts consider whether it qualifies as a nuisance. Although the Court of Appeal did not think that the argument applied in the GMF case, it did agree with it generally. In the absence of special circumstances, it explained, 'complaints with respect to nuisances created by agricultural operations should generally be brought first before the Normal Farm Practices Board before any action in nuisance is entertained by the courts.'[16]

The court action against GMF did not put the matter to rest at the NFPPB, which pursued its own investigation into the farm's practices – an investigation that it had yet to conclude by May 2006, the time of this writing. In the past, the board has shown considerable sympathy for similar operations. In a 1999 decision concerning the Mushroom Producers' Co-operative in Burford Township, the board concluded that conventional substrate composting was consistent with normal farm practices. It acknowledged that the process created extremely unpleasant odours – odours so unpleasant that they made some nearby residents vomit. It expressed its sympathy, saying that 'there is no doubt that their enjoyment of life has been substantially diminished.'[17]

The board even admitted that far less offensive methods of substrate production existed. Expert witnesses appearing before it had reported on superior methods in use in three other provinces and in several European countries. They had described the use of aerated floors to reduce the odours resulting from anaerobic reactions – a technology mandated in British Columbia – and the installation of biofilters to treat any odours that were produced. But the board concluded that the use of such alternatives remained at an experimental stage in Ontario, and could not yet be considered custom or standard. Because conventional composting methods remained standard in the industry, they would be protected as 'normal practices' under the FFPPA.

The board arrived at this decision 'somewhat reluctantly,' commenting, 'It is unfortunate that the mushroom industry in Ontario appears to be hesitant with regard to the development of that technology.'[18] It strongly urged the industry to expend the money necessary to develop technologies to reduce odours, and warned that it might be less lenient in the future. It seemed oddly unaware that its decision, replete with assurances that current methods were acceptable, provided little incentive for the kinds of changes it envisioned.

It was against this background that the NFPPB began considering the GMF case. In a decision issued during the interlude between the rulings of Justice Ferguson and the Court of Appeal, the board openly challenged the trial judge's approach. It expressed concern that he had not had the opportunity to consider its decision in the Burford Mushroom Producers' Co-operative case and that he 'did not fully appreciate the approach generally taken by the Board.' It insisted that it was not bound by his decision regarding normal farm practices, and that it could hear evidence and arrive at its own conclusions. It hinted that its conclusion might be very different: 'even though normal farm practices may cause "discomfort and inconvenience" to other persons, those discomforts and inconveniences are the price which may have to be paid if the Province of Ontario chooses to maintain viable agricultural businesses.'[19]

Five years later, the NFPPB had still not ruled on GMF's practices.[20] History was of little assistance in predicting the board's decision. On one hand, the board had ruled that the similar – and even less responsible – practices of the Burford Mushroom Producers' Co-operative were normal. On the other hand, two courts had determined that GMF's practices were not normal, and the Supreme Court had denied further appeals. Board decisions would be subject to appeal to the Divisional Court, which would in turn be subject to appeal to the Court of Appeal – which had already stated its opinion on the matter.

Seemingly oblivious to the absurdity of its position, the NFPPB asked GMF to retain a consultant to measure odours from its operations and to compare them to those produced at a hog farm. (The board refused comment on its reasons for requesting such a comparison. One can only hope that it has not deemed hog farm odours the 'new normal.') GMF's neighbours, lacking confidence in the timing of the consultant's tests and the validity of any comparisons it may make, hired their own consultant – which documented odours at levels more than 300 times those permitted by the Ministry of Environment in several other industries.

Concerns that odours of such magnitude violated not just the common law but also Ontario's Environmental Protection Act (EPA) prompted one of GMF's neighbours to launch a private prosecution in parallel with the civil action.[21] The relationship between the FFPPA and the EPA is somewhat confused. The former clearly states that it is subject to the latter. The latter, however, includes numerous exemptions for farmers following normal practices – without defining normal. The government has said that it will not prosecute a farmer under the EPA until the NFPPB has decided if his practice is normal.[22] In short, the board set up under the FFPPA effectively determines which prosecutions can proceed under the EPA – the very act to which the FFPPA is supposed to be subject.

With the regulatory dog chasing its own tail, running in ever faster circles and blurring the lines of accountability, conflicts between farmers and their neighbours are inevitable. And indeed, the complaints against GMF are by no means unique. A number of Ontario's mushroom farms face odour complaints from nearby residents. All such operations and their neighbours would be justified in wondering where they stand. Both sides can legitimately demand greater clarity and predictability from the laws and regulations governing them. Farmers must be able to predict their costs of doing business and make appropriate investment decisions. And rural residents must have assurances that a new mushroom farm in the neighbourhood won't make them sick or destroy their property values.

Mushroom composting is by no means the only controversial farming practice in Ontario. Nor is it by any means the only practice examined by the NFPPB. Since its inception, the board has heard complaints about a wide variety of issues. It has considered odours from hog, cattle, and poultry farms and from the spreading of vegetable wastes on crops. It has examined noises generated by irrigation equipment, barn ventilators, a diesel tractor, propane cannons, a wind turbine, and cowbells. It has evaluated dust associated with the cultivation of ginseng, the drying and storage of grain, vegetable farming on muck soils, and ventilation and manure removal at a poultry farm. It has dealt with flies from chicken, cattle, and hog farms, light from a greenhouse, vibrations from truck traffic, and smoke from the burning of manure – all the time, with an eye to whether the disputed practice qualified as normal.

The board has also deliberated on several by-laws. It has heard complaints about by-laws capping the number of livestock on a farm, calling for ownership of specified amounts of land for manure spreading, requiring a site plan before granting a permit for a new poultry barn, increasing minimum distance separations, prohibiting the use of tire fencing, and forbidding the burning of trees.

As is typical of the laws and regulations governing agriculture in Ontario, considerable confusion surrounds the NFPPB's mandate. The NFPPB's reach is ambiguous. Indeed, its jurisdiction has been the subject of more than one court challenge.[23] The agriculture ministry states that the nuisances under the board's jurisdiction do not include activities that could be harmful or dangerous to people or the environment.[24] It suggests that the board governs odours that are insufficiently severe to create health problems; severe odours, it says, fall under the Environmental Protection Act.[25] Regardless, the board has approved several activities that have made neighbours physically ill. The Burford Mushroom Producers' Co-operative case described above is but one such example. Farming practices that caused physical illness were also the subject of *Dietz v. Bigras*, a case decided by the board in 1997.[26] That dispute involved clouds of black dust blowing from the fine muck soils of a drained marsh. The board determined that the blowing dust resulted from normal farm practices. In its decision, it noted that the applicants had produced evidence suggesting that the dust was hazardous to their health. One neighbour·suffered attacks of severe congestion following dust storms. Another neighbour, who had had dust samples tested by a laboratory, filed as an exhibit a Health Canada publication indicating that some of the dust particles would be hazardous when inhaled. Although the board called such concerns 'genuine,' it found that it was not necessary to make a conclusive determination as to whether the dust was hazardous to human health. Its job was to determine not if the disputed farming practices were safe but if they were normal.

Likewise, despite the agriculture ministry's claims that the FFPPA 'does not give farmers the right to pollute' and that 'nuisance issues do not include activities that could be harmful or dangerous to ... the environment,'[27] the NFPPB has approved practices that threaten ground and surface waters. Although it has no mandate to hear disputes concerning water contamination, it has, in reviewing odours, dust, noise, and other nuisances, approved practices that have increased the likelihood of such pollution.

In *Knip v. Township of Biddulph*, the board considered a hog farmer's request not to be bound by a by-law limiting the size and density of livestock operations.[28] The farmer wanted instead to abide by a nutrient management plan tailored to his farm. In its 1998 decision – which was overturned on appeal to the Divisional Court – the board noted that the disputed by-law was intended to prevent environmental problems that could arise if large quantities of manure were improperly managed by farmers. It acknowledged that citizens were 'legitimately concerned' that manure could contaminate not only the shallow aquifers from which the majority of local wells drew their water but also the deeper aquifer that mixed in places with the shallow ones. The board recognized that intensive livestock operations, if not carefully managed, could have a significant negative effect upon abutting lands and could also affect the drinking water and environment of those who lived a considerable distance away.

Regardless of these concerns about safety, the board determined that capping the number of livestock units that could be located on one property restricted normal farm practices. It also determined that a nutrient management plan was a normal farm practice, and that the use of a properly monitored plan should provide for the appropriate management of manure generated by large livestock facilities. It was not swayed by the community's concern that some farmers would not follow nutrient management plans and that they would not be forced to do so. A witness from the Ministry of Agriculture pointed out that 'no one wants to be the "manure police."' A citizen complained that the Ministry of Environment was unlikely to investigate pollution incidents. The board acknowledged, 'This is not the first hearing in which this Board has heard witnesses complain that the MOE is less than diligent in the investigation of pollution running off farm properties.' Nonetheless, the board's decision – until overturned – meant that the township could not avoid the enforcement problem by prohibiting intensive farming operations.

In *Kelly v. Alderman*, the NFPPB also addressed manure management practices that could contaminate water.[29] The dispute involved, in part, the spreading of liquid swine manure. A farmer's neighbour produced pictures showing water flowing off the farmer's fields and into ditches – generally designed to carry excess water from fields to nearby streams – during and after the spreading of manure in the winter. The board admitted that it was logical to infer that some manure would have flowed into the ditches after winter spreading. More generally, it noted that the farmer's no-till cropping practices increased the potential for manure run-off. However, it stressed that water pollution was beyond its jurisdiction. In its consideration of manure spreading practices, it would restrict itself to the issue of odour. The board made no ruling regarding winter spreading. On the question of no-till cropping, it ruled that although the practice did not provide the best odour reduction, it would be considered normal as long as it occurred beyond a specified distance from nearby homes. In short, the board approved a practice that greatly increased the likelihood of manure run-off.

However ambiguous the board's mandate may be, one thing is clear: the board has followed the direction established in the preamble to the Farming and Food Production Protection Act. It has conserved, protected, and encouraged agriculture. Justice, as meted out by the NFPPB, has by no means been blind. As Ontario Superior Court Justice Gordon Killeen wrote, 'A careful reading of this entire Act must lead to the conclusion that the balancing reflected in it tends to give the benefit of a large doubt or edge to the farmer, big or small as the case may be.'[30] The board's tendency to give the benefit of a large doubt to the farmer contrasts with the disinterested common-law rules governing nuisances – rules that, rather than favouring one party over another or picking winning and losing occupations, let the chips fall where they may.

The tendency to favour one class of rural residents over another is not the only characteristic that sets the NFPPB's decisions apart from those issued by the courts under the common law. Board decisions tend to be somewhat less predictable and less clear-cut than those issued by their common-law predecessors. Because the NFPPB looks at complaints on a case-by-case basis, there is no clear pattern to the forty–plus decisions it has recorded to date. Sometimes it has defended a nuisance in one situation and ruled against a similar nuisance in another situation. For example, although it decided – twice – that dust blowing from the muck soils of vegetable farms north of Point Pelee resulted from normal practices, it decided that dust blowing from the sandy soils prepared for a ginseng crop did not result from normal practices.[31] Likewise, the board has sided both with and against the use of propane-fired cannons to scare wildlife away from crops. It ruled that using a 'bird banger' throughout the night to protect a farmer's vegetables from deer was not normal. However, in two other cases, it ruled that the nighttime use of such a device to protect sweet corn from racoons was normal, as was its daytime use in a graperie.[32]

Despite the absence of predictable patterns, several characteristics set the NFPPB's decisions apart from those that courts tend to issue under the common law. One difference appears in the board's tendency to compromise. Indeed, rural planner Wayne Caldwell identifies the board's ability to recommend compromises as its biggest difference with 'the win/lose character of the court system.'[33] In some cases, the board's orders reflect compromises to which all parties have agreed. Typical was its order in a case concerning smells arising from the incineration of dead hens at a poultry farm. The board ordered the farmers to freeze dead birds and to incinerate them no more than three times a month; it also forbade the farmers to incinerate when the wind blew from west to east.[34] In another compromise, neighbours agreed to share in the cost of dismantling and moving a noisy wind turbine.[35]

Sometimes such compromises make it difficult to determine who 'won' or 'lost' in front of the board. For example, although the board described its ruling in a dispute about noise from cowbells as favouring the farmer, it in fact ordered the farmer to construct a fence that would keep his belled cows away from his neighbour's home during the night.[36] In a dispute over odours from a swine nursery, the board maintained that it ruled in favour of the applicant; however, it found that the hog barn and manure storage facility complained of conformed to normal practices and it permitted manure spreading practices that the applicants feared contaminated local waters.[37] In another case in which the board claimed to have ruled in favour of the applicants, it issued an order to the farmer that failed to satisfy the applicants, who feared that creating a five-foot earth berm and planting trees on it would not sufficiently reduce the noise and dust created by the ventilation and cleaning of the farmer's poultry barns.[38]

Another difference between the NFPPB and the courts is found in the former's tendency to micromanage solutions. In a decision concerning noise from a tractor-driven irrigation pump, the board specified the precise location and design of a noise barrier to be erected by the farmer. It insisted that the barrier be constructed of three-quarter-inch plywood lined with four-inch-thick fiberglass insulation. Furthermore, it limited the pump's hours of operation, required the farmer to notify his neighbours seventy-two hours before irrigating, and called for weekly readings of noise levels – upon receipt of which it would judge the sufficiency of said alterations.[39] In a decision concerning blowing soil and sand, the board ordered the farmer to correct the nutrients in his soil, to seed his land with a small grain and to underseed specific areas with clover. It also ordered the farmer to install windbreak netting and to plant trees as permanent windbreaks.[40] In a decision concerning noise and dust resulting from a farmer's grain handling and storage, the board ordered several specific solutions, including the construction of a concrete block building around one fan and the use of foam-insulated fan covers to further buffer the noise. It also recommended remedial work to reduce dust, suggesting a variety of measures that would have the effect of enclosing operations.[41] The board's tendency to micromanage farm practices – a tendency more typical of central planners than judges – reflects a lack of confidence in farmers. Indeed, the FFPPA rests on the assumption that farmers need help in solving their own problems. As Justice Killeen said of the FFPPA's preamble, 'The extensive recitals to this Act reflect the somewhat protective and paternalistic position adopted by the government towards the agricultural community.'[42]

Perhaps the most important distinction between the NFPPB and the courts is found in the former's focus on *inputs* rather than *outcomes* – its focus on a farmer's activities rather than on those activities' impacts on others. The board's obligation is to determine whether a practice is normal. This contrasts sharply with a court's traditional obligation, which is to determine whether a practice unreasonably harms others. Harm is of little import to the board. A decision in favour of a dairy farmer's manure storage practices typified the board's thinking on this matter. Although the board found that the farmer's neighbours experienced 'substantial inconvenience' from the odours associated with the farm, it concluded that because a significant portion of the odours resulted from normal farm practices, the farmer was entitled to continue to utilize those practices.[43] As the board so succinctly explained in another case, 'Normal farm practices are protected and may continue even if they cause the Applicants to be aggrieved.'[44]

Allowing one person's activities to harm another is anathema to the courts. Justice Ferguson, in his 1999 decision regarding Greenwood Mushroom Farms, aptly summed up the injustice inherent in right-to-farm legislation: 'I must say that I am troubled by a policy that a farmer can cause serious harm to a neighbour as the result of a normal farm practice without that neighbour having any remedy in damages. From the plaintiffs' point of view it does not seem just that the neighbour should suffer a serious loss without compensation in order that the whole community can benefit from the production of agricultural products.'[45]

Just or unjust, Ontario's farmers continue to push for even greater protection. In 2004, they impressed upon the provincial government's

Agricultural Advisory Team their wish for a stronger Farm and Food Production Protection Act.[46] The government may be all-too-inclined to heed their demands. That it remains in the thrall of the industry was suggested in Premier Dalton McGuinty's remarks upon announcing new financial assistance for farmers: 'We're taking this step because we know how important the agriculture sector is to the economic and social well-being of every person in this province.' Then Agriculture Minister Steve Peters added, 'This is a government that is committed to rural Ontario.'[47] Non-farming rural residents could be forgiven for disagreeing.

6 Beyond the Right to Farm: Changing Drainage and Planning Laws to Minimize Restraints on Farming

I can't advise farm business people to uncritically endorse this remedy. Think of the consequences, should private property rights be over-emphasized.

– Elbert van Donkersgoed, 2005[1]

The Right to Drain

In March 2005, Elbert van Donkersgoed counselled farmers against endorsing stronger property rights. Then adviser to the Christian Farmers Federation of Ontario, long time advocate of the family farm, and author of a weekly column on agricultural issues, Mr Donkersgoed was alarmed by rural landowners' suggestions that private property rights should balance burgeoning government regulations. Stronger property rights, he warned, would not only limit farmers' freedom to cause discomfort from nuisance, dust, and odour but would also undermine the legislation that supports the creation of drainage systems. 'Enshrined property rights,' he explained, 'would make it much harder to create and maintain these multiple-property drainage projects.'

While the right-to-farm statutes that have superseded the common law are perhaps the clearest examples of provincial governments overriding the property rights of those who are harmed by agriculture, they are neither the only nor the first examples. Laws governing the drainage of agricultural land long ago replaced the common law – and its delicate balance of property rights – with a complex administrative system overseen by provincially appointed decision makers.

Provincial governments have wanted to facilitate drainage for obvious reasons. As the Ontario Ministry of Agriculture and Food explains,

'Profitable returns from farmland depend on effective drainage.'[2] The most frequently cited benefits include improved field conditions that advance planting dates and accommodate heavy equipment, increased crop yields, and more valuable land. The Land Improvement Contractors of Ontario elaborate on the benefits as follows: 'The object is to improve the productivity of the farmers' land base, reduce the unit cost of production, provide for a wider range of crops that may be grown, reduce soil compaction, reduce surface runoff and limit the surface erosion of fields ... Drained land permits the farmer to grow the same amount of produce on 40 percent less land. It also reduces the inputs needed for efficient crop production.'[3]

Such benefits, however, do not come without costs. Agricultural drainage systems have led to significant losses of wetlands, which store run-off, cleanse waters, and provide wildlife habitat. Agricultural drains carry sediment to lakes and rivers. With the drained waters and sediments flow a number of pollutants. Sometimes referred to as 'nutrient highways,' surface drains can transport nitrogen and phosphorus from farmers' fields to nearby rivers and lakes. These open ditches can also carry bacteria and agricultural chemicals into waterways. Subsurface drains, too, can act as conduits for nutrients, pesticides, and pathogens. Such drains, traditionally made of clay or concrete cylindrical tiles, are now generally made of perforated plastic tubing. Typically, pollutants from liquid manure spread on un-tilled land travel to subsurface drains through macropores – cracks, worm holes, or root channels – in the soil. The transport can be very rapid, with drain outlets discharging pollutants within twenty minutes of the spreading of manure.[4]

Recent charges laid by Ontario's Ministry of Environment illustrate a few of the many ways that agricultural drainage can threaten water quality. In one case, after the operators of a hog farm spread liquid manure on frozen ground, foul-smelling effluent rushed through an open ditch towards a nearby lake. In another, manure applied during a rain storm made its way into subsurface drains. In another, manure from an accidentally uncapped hose spilled into a drain. And in yet another, a dairy farmer excavated a channel in order to divert field runoff from his barn into a municipal ditch. More frequently, pollution from drains goes unnoticed, both by those who create it and by those responsible for regulating it. As explained in an Agriculture and Agri-Food Canada report, 'Too often it seems that drainage water is forgotten once it leaves the drained area.'[5]

The common law, where still in force, provides a mechanism for regulating otherwise 'forgotten' drainage waters. It limits a farmer's right to drain his land and holds him accountable for the damage resulting from many drains. The common law distinguishes between water draining from land through a natural watercourse and surface water draining from land through no defined course. An upper landowner has a right to drain lands through a natural watercourse. Likewise, a lower landowner has an obligation to accept waters so drained. In the words of the Ontario Ministry of Agriculture and Food, 'If water is in a natural watercourse, it must be permitted to flow.'[6]

On the other hand, a landowner has no right to drain surface water that does not flow through a natural watercourse. He has no right to collect rainwater in subsurface drains or ditches and send it onto his neighbour's land. Conversely, his neighbour has no obligation to receive the drainage of such surface water. He may build a berm or dyke around his land to prevent it from intruding, and he may refuse his neighbour permission to run a drain across his property or to tile into his own drainage system.

In a 1934 decision upheld by the Supreme Court of Canada, the Saskatchewan Court of Appeal spelled out the rules that existed under the common law: 'The upper proprietor has no legal right, as an incident of his estate, to have surface water falling on his land discharged on the lower estate, although it naturally would find its way there.' Conversely, 'a lower proprietor owes no duty to an upper proprietor to receive the surface water naturally coming from the upper land.' Accordingly, 'the lower proprietor may lawfully, in the proper use of his land, erect obstructions to prevent the water from overrunning his land, even if such obstruction has the effect of forcing the water back on the lands of the upper proprietor.'[7]

The court noted that this rule follows logically from the fact that, under the common law, a lower proprietor has no right to the flow of surface waters. He cannot demand access to waters that would come naturally onto his property unless they flow in defined channels or watercourses. 'If then an upper proprietor may deal with surface water as he pleases and owes no duty to a lower proprietor to allow the water to flow to his land, which would naturally go there, it would appear most unreasonable that the lower proprietor must receive the water when he does not want it but has no right to it when perhaps he wants it.'[8]

While citing this decision approvingly, the judge in a 1973 Nova Scotia case regarding blocked drainage applied the law somewhat differently. In the latter case, the defendant raised the level of its land and blocked a natural drain that was not a watercourse. The judge determined that the defendant had drained the waters on its own property and cast them onto the plaintiff's property. This constituted a nuisance. 'A party cannot,' the judge explained, 'by artificial means gather the water on his property and throw it upon his neighbour's land.' As the Ontario Court of Appeal had found in 1950, 'the defendant was not entitled to send on to the plaintiff's land any surface water which would not have reached there naturally.'[9] Or, as had been more recently confirmed in a Saskatchewan decision, 'An owner of land has no right to rid his land of surface water, or superficially percolating water, by collecting it in artificial channels and discharging it through or upon the land of an adjoining proprietor.'[10]

Working within these well-established common-law rules, neighbours may negotiate mutually agreeable solutions to drainage problems. They may construct what in Ontario are known as 'mutual agreement drains' under agreements spelling out each party's responsibilities for financing, constructing, and maintaining the drain.

Legislation overriding the common law has, however, long enabled many landowners to avoid negotiating with their neighbours. In what is now Ontario, relief came as early as 1858, with An Act respecting the Municipal Institutions of Upper Canada. The act allowed a majority of the owners of the property in any part of a township to petition their council for the drainage of the locality. The council could then hire an engineer to plan the work and pass a by-law providing for the work's construction and financing through assessments charged to all properties benefiting from it.

Today, under Ontario's Drainage Act, landowners may still petition their local council for the construction of a drain across others' properties. Once a majority of affected landowners or owners holding 60 per cent of the affected land sign the petition, and a municipally appointed drainage engineer designs the drain, the council may commence work on the requested drain and assess local residents for their share of the costs – which are often artificially low, thanks to generous provincial loans and grants.[11] Although the elaborate planning system allows for individuals to express concerns at public meetings and through formal appeals, the problem is generally approached as one of efficiency and the common good – notions frequently at the heart of administrative decisions about public utilities but rarely contemplated under the traditional common law.

Fisheries and Oceans Canada describes Ontario's Drainage Act as legislation that 'balances the rights of landowners living along water-courses with the rights of property owners who do not have access to a stream or creek in order to drain their lands.'[12] This description ignores one essential factor: the act and its predecessors have not balanced pre-existing rights. Instead, they have created new rights for those without access to streams and curtailed the earlier rights of their neighbours not to receive their surface waters. In doing so, they have redistributed, rather than balanced, rights.

The acts have doubtless facilitated drainage. Ontario is Canada's most extensively drained province, with more than 45 per cent of its cultivated land drained by artificial means.[13] But the administrative regime created by the act has not been able to ensure that drainage proceeds smoothly. According to the Ontario Ministry of Agriculture and Food, 'drainage of water is one [of] the most common areas of dispute between rural neighbours.'[14] One such dispute has involved a drain discharging into the Blanche River, near New Liskeard, Ontario. The drain has carved a ravine that is several hundred feet long, 100 feet wide, and more than 15 feet deep, washing thousands of tonnes of sediment into the Blanche River. Downstream landowners have objected – to no avail – to provisions in the Drainage Act that sanction the erosion of their land. If the drainage engineer determines that the costs of establishing a sufficient outlet for a drain are prohibitive, he may approve a design that will injure downstream lands as long as he determines appropriate compensation for the injured landowners.[15]

One of the province's most infamous drains has been the Ramara Township's McNabb Drain, described by one environment official as 'a dysfunctional and unstable system.' After its expansion in 1997 – carried out without consultation with the downstream landowners who would ultimately suffer the nuisances it created, and, until 2004, ignored by the environment ministry – the drain has frequently discharged waters laden with sediment, nutrients, and bacteria into a natural watercourse and, from there, into Lake Simcoe. Water resources engineer Ted Cooper, representing downstream resort and farm owners, blames the drain for as many as fifty incidents of erosion or pollution between 1998 and 2003. Mr Cooper has harsh words for the drainage regime that has supplanted the common law, concluding that it 'has had a devastating impact on the environment, and [on] the water resources of the Province.'[16]

Drainage works do not just benefit from exemptions from the common law. They also get special treatment under Ontario's Environmental Assessment Act. The act requires municipalities and public bodies to obtain environmental approvals for a wide range of public-sector undertakings, such as roads, waste management facilities, sewage works, and flood protection works. Section 5 of the act specifies that 'every proponent who wishes to proceed with an undertaking shall apply to the Minister for approval to do so.' The proponent must, after consultation with interested parties, submit terms of reference and, subsequently, an environmental assessment for approval. Regulation 334 exempts from these requirements drainage works regulated under the Drainage Act. While not without scrutiny – local conservation authorities and the Ministry of Natural Resources review proposed drains – drainage works also evade other commonly required approvals. Since 1962, they have been exempt from the requirements, under the Ontario Water Resources Commission Act and its successor, the Ontario Water Resources Act, that sewage works be approved by the Water Resources Commission or, more recently, the Ministry of the Environment.[17]

Planning for Agriculture

Just as legislators have moved decisions about drainage from individuals to majorities, and from common-law courts to bureaucrats, so too, in a relentless trend towards centralized decision making, have they moved decisions about planning from municipalities to the provinces themselves. Nowhere is this better illustrated than in Alberta, where the provincial government has systematically reduced municipal control over intensive livestock operations. It has done so to overcome local opposition to agricultural expansion, such as that experienced by the Taiwan Sugar Corporation.

In the late 1990s, the state-owned TSC faced growing opposition at home. The first company in Taiwan to promote industrial farming, it raised not only 100,000 hectares of sugarcane but also approximately 400,000 pigs. It was finding, one company representative explained, that 'mounting environmental concern and criticism in recent years has made it hard to raise hogs in Taiwan.' With the blessing of Taiwan's parliament, TSC decided to invest abroad. Canada's abundant feed grain and open spaces beckoned. 'Canada is such a huge country,' TSC reasoned, 'and we can easily find a place without residents nearby.'[18]

Attracted, doubtless, by the Alberta government's promise of minimal oversight and its vow to 'remove as many regulations as possible,'[19] TSC assembled proposals to produce between 80,000 and 150,000 hogs annually at several sites in one of two Alberta communities – Foremost, in the County of Forty Mile, and Hardisty, in Flagstaff County. In 2000, the planning commission of the County of Forty Mile, reflecting local concerns about odours and potential water contamination, unanimously rejected TSC's application. In contrast, Flagstaff County, despite receiving more than 160 letters and a 2000–name petition in opposition to the project, granted TSC a permit to construct its complex. Hardisty residents sued. In 2002, the Alberta Court of Appeal overturned the development permit for the proposed farm.

Taiwan Sugar Corporation's foiled efforts called attention to a growing problem for industrial farms: communities were increasingly resisting their overtures. Initially, the province respected the right of municipalities to restrict farming. Indeed, in 1998, Premier Ralph Klein insisted that 'it would be inappropriate for the province to intervene' in municipal planning and development issues.[20] But the province's patience soon wore thin. In 2001, made wary by the effective local opposition to TSC and other livestock producers, Mr Klein's Progressive Conservative government introduced legislation to transfer responsibility for the siting and monitoring of intensive livestock operations from municipalities to a provincially appointed board.

The legislation was not Alberta's first attempt to force municipalities to accommodate intensive farms. In 1999, the province had amended the Municipal Government Act to require municipalities to address the protection of agricultural operations in both their development plans and their land-use by-laws. Minister of Municipal Affairs Iris Evans explained at the time that the amendments would 'require municipalities to recognize the importance of agricultural operations as a valuable resource in this province.' Her colleague Gary Severtson was more direct: the amendments, he noted approvingly, were about 'ensur[ing] that municipalities cannot use by-laws to limit normal farming practices.'[21]

The 1999 amendments failed to provide the protections that legislators sought. In 2000, the agriculture minister appointed the Sustainable Management of the Livestock Industry in Alberta Committee, comprised of three MLAs, a municipal representative, and a representative of the cattle industry, to advise him on the development and operation of the province's livestock industry. In its April 2001 report, the committee pointed out that municipalities were increasingly tightening their land-use rules, requiring more detailed information from live-stock producers, and introducing environmental protection standards and other operational restrictions.[22] The committee recommended a provincial approval process for intensive livestock operations. A provincially appointed board, it suggested, should make final decisions on land use and technical requirements related to new and expanding operations, ensuring consistent, province-wide standards along with greater certainty for the industry.

In November 2001, citing the livestock industry's economic importance, the government introduced amendments to the Agricultural Operation Practices Act. Under the amended act, the provincial Natural Resources Conservation Board would 'provide a one-window approach for the livestock industry and the public.'[23] Like many other right-to-farm laws, the amended AOPA attempted to move disputes between farmers and their neighbours from the common law to provincially appointed regulators – the NRCB, the Farmers' Advocate, and an Agricultural Practices Review Committee largely comprised of industry peers. But the amended act carried centralization further: it also moved decisions about the siting and regulation of new and expanding confined feeding operations from municipal and county officials to the NRCB. While affected individuals and municipalities would be encouraged to provide input into NRCB processes, the decisions would no longer be their own. As MLA Albert Klapstein explained, 'in the end, when push comes to shove, the NRCB can decide and can in fact over-rule a municipality's position.'[24]

The provincial government offered a number of reasons for wresting control from municipalities. It complained that decisions by myriad municipal councils lacked consistency. 'In order for our livestock industry to grow,' it maintained, 'we need to eliminate uncertainty and inconsistency.' It also argued that some municipalities lacked technical expertise while others lacked the will to deal with conflict and that, as a result, municipal decisions were sometimes based on 'emotion and political expediency.'[25]

Both opposition parties raised red flags about the plan to disempower municipalities. Liberal MLA Bill Bonner identified one of the few weaknesses of the bill as 'the fact that the decisions that impact ... people are not going to be made at a local level and not by people who are familiar with that particular region.'[26] Fellow Liberal MLA Ken Nicol lamented, 'we've really taken away from those local communities the chance to determine who they are, what they are, [and] the kind of economic activity that goes on in that community with respect to live-stock production ... Under this bill the local community cannot say no.' Regardless, Mr Nicol did not suggest giving local communities a veto over all industrial agriculture. He proposed instead that communities be required to designate some land on which confined feeding opera-tions would be permitted. 'I truly believe,' he stressed, 'that we have to also make sure that a county doesn't have the option to say: zero, no confined feeding operation.'[27]

The NDP expressed less ambiguous concerns about reducing munic-ipal authority. MLA Brian Mason identified the change as a part of the government's strategy to extend unpopular farming operations:

> In our view, it is a way of restricting the ability of a rural municipality to prevent large-scale hog operations in particular from being placed in their county in places that may have an impact on the surrounding neighbours. We believe that municipal government in this province is fully competent to make these decisions, that [they are] better made by those people who have to live next to the operation than by a board or a bureaucrat in Edm-onton, and that we should respect municipal autonomy ... What we have now is at least that people who have to live with the decision can make the decision. They may not make it right every time, but they're the ones that we should be vesting this authority in.[28]

As Mr Mason had earlier explained, 'That is exactly what local munici-pal democracy is all about. If the people don't want it, who is going to say that they're wrong and they should be overruled?'[29] Mr Mason warned his colleagues that people would not look favourably on a gov-ernment that did away with local autonomy:

> The people will recognize that the smell goes far beyond just the livestock operations, that it extends as well to a government that puts the hog industry ahead of the quality of life of the people of Alberta ... [It] is going to be a very, very serious problem for all the members of this Assembly to deal with over the coming years as these operations multiply and as the provincial government does away with local autonomy in order to facili-tate their development. The people will know where the smell is coming from.[30]

Opposition concerns went unheeded, and the changes took effect in January 2002. At least one group of farmers concerned about their changing industry laments the loss of local control. The Society for Environmentally Responsible Livestock Operations argues that, before 2002, local officials had both the tools and the incentives to make sustainable decisions: they knew their environments intimately, worked with guidelines that were sufficiently flexible to suit local conditions, were accountable to those who elected them, and, as members of the community, had to live with the effects of their decisions. In contrast, distant officials, tellingly called 'Approval Officers,' tend to rubber stamp applications in a biased process that silences and alienates concerned neighbours. The NRCB, SERLO charges, provides 'a path of least resistance for the livestock industry to expand.' The board has approved applications that local governments have rejected, has failed to enforce compliance with conditions of approval, has removed conditions that local governments have placed on farms, and has dismissed complaints that local communities have expressed about existing livestock operations. 'Stripping away local control,' SERLO concludes, 'leads to the fast tracking of livestock operations, improper siting, and the destruction of community cohesiveness.'[31]

Alberta is by no means the only province to promote agriculture by stripping away municipal control over the siting and operation of farms. When Ontario municipalities began throwing up barriers to intensive livestock farms, provincial legislators there, too, passed a series of laws freeing farmers from municipal restraints.

At one time, Ontario's municipal affairs ministry encouraged municipal independence. In 1997, planning and policy director Bryan Hill objected to the agriculture ministry's suggestion that there was a need to protect farmers from municipalities. Citing the 'fundamental principle that councils are duly elected and responsible governments,' he explained that 'municipalities have the local autonomy to make, and be accountable for, decisions which are based on the interests of the entire community.' He added, 'the municipality is the most appropriate level of government to determine the local regulations that should apply' to farming.[32]

Ontario municipalities traditionally exercised their land-use control powers through building-permit processes requiring compliance with zoning by-laws and minimum distance separation rules. In the last decade, an increasing number of municipalities required prospective farmers to produce manure management plans. Some went further, limiting the number of livestock – or even the number of employees – allowed on farms, specifying a minimum capacity for storing manure or a minimum land base for spreading it, demanding studies of prevailing winds and road traffic, or requiring public meetings to review proposals.

Such by-laws, numbering approximately eighty-five by 2003,[33] proved unpopular with the provincial government. In 2000, the agriculture ministry described as a 'problem' the restrictive municipal response to proposed farms.[34] Progressive Conservative leadership contender and future premier of Ontario Ernie Eves accused municipalities of being 'offside' in passing manure management by-laws that 'may go beyond what we want to do.'[35] The province fought individual municipalities, challenging several by-laws at the Ontario Municipal Board.[36] But it also took a more systematic approach, starting in 1998 with revisions to the provincial right-to-farm law.

The 1998 Farming and Food Production Protection Act freed farmers from the need to comply with municipal by-laws that restrict normal farm practices: 'No municipal by-law applies to restrict a normal farm practice carried on as part of an agricultural operation.' Since the passage of the act, farmers have turned to the Normal Farm Practices Protection Board to challenge the application of a variety of municipal by-laws, including several capping the number of animals permitted on a farm or requiring the ownership or long-term control of a specified amount of land for spreading manure. In several cases, the board has found that such by-laws did, indeed, restrict normal practices and should not apply to the farmer in question. However, a case currently before the courts suggests that the board may have overstepped its authority to free farmers from restrictive by-laws. In January 2005, the Divisional Court ruled that although the board may exempt normal farm practices from municipal by-laws prohibiting or regulating nuisances, it has no jurisdiction to hear cases involving zoning by-laws governing land-use planning.[37]

Although the 1998 reforms to the Farming and Food Production Protection Act did much to shield individual farmers from some municipal by-laws, they did nothing to overturn the by-laws themselves. The Progressive Conservatives addressed that issue in 2001 by introducing Bill 81, the Nutrient Management Act, which would establish provincewide standards for the management of manure and other nutrient-rich materials.[38] The act passed the following year and came into force in July 2003; the general regulation under the act came into force three months later. At first, the regulation applied only to new and expanding farms. For other farms, the rules were to be phased in over a number of years. In response to pressure from farmers, the government further postponed applying the regulation. It moved the compliance deadline for approximately 1,100 larger farms from 2004 to 2005 and agreed not to apply the regulation to smaller farms until at least 2008.

The new regime greatly reduced municipal authority to regulate manure management practices. The voluminous and far-reaching regulations made under the Nutrient Management Act purported to supersede municipal by-laws addressing the same subject matter. Despite some controversy over the extent of the supersedence, such was clearly one of the purposes of the act.[39] As agriculture minister Brian Coburn explained, by superseding municipal by-laws, the legislation would ensure 'a consistent approach and a clearly articulated set of common goals right across Ontario.'[40] The ministry offered the following illustration of the act's taking precedence over a by-law:

> A municipal by-law may not enhance or supplement the requirements of the Regulation if the substance and purpose of the by-law are the same as the Regulation. For instance, in order to protect groundwater, a municipal by-law might provide that a livestock operation shall have a nutrient storage facility that is capable of containing at least all of the nutrients generated or received in the course of the operation during a period of 300 days. However, the Regulation may be seen to cover the same topic as it provides that a livestock operation need only have a nutrient storage facility that is capable of containing at least all of the nutrients generated or received in the course of the operation during a period of 240 days. This provision of the by-law also has the same purpose as the Regulation in that they are both designed to protect the natural environment. As a result, this provision of the Regulation supersedes the by-law as both provisions address the same substance (i.e. nutrient storage capacity) for the same purpose (i.e. protection of the natural environment).[41]

Bill 81's provisions for municipal disempowerment received only a muted reaction. The NDP's Marilyn Churley was the only MPP to focus on the issue when the bill was being debated in the Legislative Assembly or discussed in committee hearings.[42] She expressed concern that uniform standards might not be sufficient, especially in environmentally sensitive areas. 'You can have different situations in different rural areas,' she explained. 'There might be other mitigating factors within that area which would cry out for a higher level of control.'[43] Municipalities, she maintained, are best suited to establishing such controls. 'A municipality, after all, knows its own district better than we sitting on high up here. I get very concerned and very nervous when you have overreaching legislation that takes away the ability of politicians, who know their jurisdictions best and can consult with all aspects of the community, to set rules and by-laws.'[44] Ms Churley also accused the government of ignoring the 2001 Supreme Court ruling upholding a municipality's right to enact a by-law restricting pesticide use.[45]

Several environmental groups likewise opposed Bill 81's rebalancing of power. The Sierra Legal Defence Fund feared that provincial standards could be lower than those adopted by municipalities, decreasing protection of the environment and quality of life. The Supreme Court pesticide by-law ruling, it noted with approval, 'mentioned that it was important that municipalities be empowered to improve upon, but not lower, standards from higher orders of government.'[46] The Sierra Club agreed that the provincial law should require minimal protection upon which municipalities with geographical and societal peculiarities could expand, arguing, 'you have to allow municipalities to develop and to be diverse.'[47] The Canadian Institute for Environmental Law and Policy also expressed concern about eliminating the municipal role, noting that 'municipal by-laws have been one of the principal means of addressing agricultural impacts to water quality in recent years.'[48]

Farmers, on the other hand, largely supported the changes – at least initially. Ontario Pork pronounced itself 'very pleased' with the introduction of the legislation – especially the provisions replacing a 'patchwork' of municipal by-laws with province wide standards.[49] The Pork Industry Council rushed to 'congratulate' the government and to express its strong support for superseding municipal by-laws.[50] The Ontario Federation of Agriculture 'applauded' the introduction of the act, pleased that it would 'alleviate the inconsistencies that municipal nutrient management by-laws posed.'[51] Municipalities, it maintained, lacked the expertise to manage the issue, 'and so a lot of them have been over-restrictive.'[52] Even after the act and its regulations came into force, the federation maintained that 'most farmers prefer provincial to municipal regulation.'[53]

Although farmers soon began to chafe at the regulations, their complaints generally concerned the complexity of the rules, the costs of compliance, and who should bear them rather than the issue of where regulatory authority lay. (The provincial Liberals responded to their complaints with a promise of $20 million to help larger farms comply; their federal counterparts followed with a promise of $57 million to fund Ontario farmers' environmental stewardship activities.) At least one agricultural organization did come to oppose the regulatory shift, not because it had disempowered municipalities but because it did not go far enough. The Christian Farmers Federation regretted its earlier calls for provincial intervention. 'With hindsight,' it lamented, 'it is now easy to see that the farm community over-reacted to the few municipalities that want to be different.' The new regime, it complained, left them worse off than before: not only did it create a 'regulatory quagmire' but it also was 'still not robust enough to supersede municipal by-laws.'[54]

The federation's latter criticism was correct to the extent that municipal by-laws will continue to apply to smaller farms until the new nutrient management regulations replace them, which may happen in 2008. But the federation did not have to worry about municipal by-laws being too restrictive in the interim: in December 2003, the Liberal government introduced Bill 26, the inaptly named Strong Communities (Planning Amendment) Act, thereby ensuring that municipal by-laws would comply with provincial policies.

Bill 26 masqueraded as one that would 'empower communities to shape their own destinies' and give them 'the much needed tools to control their own planning.'[55] In several fundamental ways, however, it did just the opposite. Indeed, columnist John Barber explained, 'the thrust of the amendments is to consolidate much more power at the provincial level.'[56] The bill severely curtailed the freedom of municipal councils and planning boards. Section 3 specified that decisions affecting planning matters 'shall be consistent with policy statements' issued under the act. No longer would municipalities or agencies merely have to 'have regard to' provincial policy statements – the requirement established in the previous Planning Act. No longer would they have the flexibility to differently interpret provincial policies – a flexibility once admired by the municipal affairs ministry as 'exactly the type of local flexibility intended.'[57]

For good measure, Bill 26 likewise restricted the power of the Ontario Municipal Board, whose willingness to uphold municipal by-laws that restrict intensive agriculture displeased the province. Once the bill took effect, not only would OMB decisions, like those of municipalities, have to 'be consistent with' provincial policies, but they could also, in matters of provincial interest, be varied or rescinded by the Lieutenant-Governor in Council.[58]

Although Bill 26 predictably met with a mix of criticism and support from opposition politicians,[59] interest groups, and the public,[60] more striking was the widespread apathy regarding the issue. Most of the town hall meetings and public information sessions on the proposed planning reforms were sparsely attended. And although the Standing Committee on General Government intended to hold public hearings on the bill in Toronto, Kapuskasing, London, and Ottawa, insufficient interest caused it to cancel all but the Toronto hearings. Thus, with relatively little debate, the Strong Communities (Planning Amendment) Act received Royal Assent in November 2004.

Under the amended Planning Act, the province revised its Provincial Policy Statement. As described by municipal affairs minister John Gerretsen, the new PPS 'sets out what communities all across Ontario should look like.'[61] It laid the ground rules for land-use planning, providing policy direction to municipalities on a number of issues, including 'better protecting agriculture.'[62] Much like its predecessor, the PPS decreed that 'prime agricultural areas shall be protected for long-term use for agriculture.' In such areas, it continued, 'all types, sizes, and intensities of agricultural uses and normal farm practices shall be promoted and protected in accordance with provincial standards.'[63] New secondary uses, it specified, shall be limited in scale and shall not hinder surrounding agricultural operations. Thanks to amendments to the Planning Act, municipalities would have no choice but to 'be consistent with' the directions contained in the new PPS. Municipal disempowerment had been achieved: no longer could communities protect themselves against unwanted farms.

The latest province to centralize agricultural decision making is Manitoba, which, in June 2005, replaced its Planning Act in order to constrain municipalities' power to determine the kinds of agricultural activities permitted within their borders. In its quest to minimize conflicts between farmers and their neighbours, Manitoba had long advocated land-use planning, not only to separate farms from residential developments, thereby reducing the impacts of odours and noises, but also to put local people on notice that farming is an acceptable land use in designated areas.[64] The province had learned, however, that it could not rely on local governments to do the kind of planning it preferred – planning that favoured agriculture over other land uses. Voluntary planning at the local level, through development plans, zoning by-laws, and permitting processes, had produced a wide variety of local standards governing the location, size, and operating practices of farms, resulting in what one minister disparaged as 'a confusing regulatory environment.'[65] Although the province had issued land-use policies and implemented technical review procedures to guide local decisions, it had failed to guarantee a uniformly warm welcome for intensive livestock operations across the province.

To remedy this perceived problem, the NDP government introduced Bill 40, the Planning Amendment Act, in March 2004. The proposed act required each municipality or planning district to adopt a development plan with a livestock operation policy setting out the areas in which livestock operations could or could not be developed or expanded. Municipal siting policies would have to be 'generally consistent' with provincial guidelines. Furthermore, the government could reject a municipality's development plan by-law or approve it 'subject to any alteration or condition that the minister considers necessary or advisable.' Although the government claimed to be enhancing the power of rural municipalities, 'respecting' their choices, and accepting the 'democratic voice of the people,'[66] at least some municipalities understood its true intent. Ed Stroeder, reeve for the rural municipality of Westborne, called the bill 'a veiled effort at bypassing local authority.'[67]

Similarly, although the government claimed to be further involving the public in the planning process, Bill 40 in fact reduced opportunities for public input. Under the bill, a municipal council would forward a proposal for a large livestock operation to the minister, who would in turn refer the proposal to a government-appointed Technical Review Committee. After receiving the committee's report, the council would hold a public hearing on the proposal. Regardless of the concerns expressed at the hearings, the council could impose only very limited conditions on any approval it issued: it could require the proposal to conform with its development plan and zoning by-laws and the recommendations proposed by the Technical Review Committee, and it could require covers on manure storage facilities or shelterbelts – rows of wind-blocking trees – to reduce odours. No other restrictions on the location, type, or size of an operation, the kind of livestock raised, or manure management practices would be permitted. Hogwatch Manitoba complained that this process 'virtually eliminates the role the public can have in influencing a council decision.'[68]

Nor were farmers themselves entirely happy with the proposed legislation. The National Farmers Union, dedicated to preserving the family farm, complained that Bill 40 'removes the power of local people and RM councils to address the problems created by industrial hog factories.' Enhanced ministerial discretion, it noted, made a mockery of municipal independence. 'Municipal Councils have the right to decide, so long as they make decisions the Provincial Government of the day agrees with. But if they don't, then the Provincial Government will decide.'[69]

Other agricultural organizations expressed concerns of a very different nature. Manitoba Pork Council objected that the bill 'fails in providing either consistency across the province or predictability in land-use decisions. We will be left with a continued patch-work of requirements and regulation across Manitoba.'[70] And Keystone Agricultural Producers, Manitoba's largest farm lobby, wanted the government to put even more restrictions on municipal by-laws concerning farming.[71]

Facing criticism from all quarters, the government withdrew Bill 40 in November 2004. Environmentalists and the National Farmers Union praised the move, as did the *Winnipeg Free Press*, which opined, 'The province tried to dictate to municipalities what business they can permit, but that should be left to citizens who elect their councils and live with the results.'[72]

The celebrations were short lived. Less than six months later, the government introduced Bill 33, the Planning Act. Like Bill 40, Bill 33 was presented as one that would 'strengthen local decision making' and 'increase the flexibility of councils to respond to local concerns.'[73] But again, like its predecessor, the bill did just the opposite. The bill maintained the measures thwarting communities' efforts to mitigate the threats posed by intensive livestock operations. Municipal development plans – which the province could alter or replace at will – could not distinguish between different types of livestock operations or farming practices. Municipalities could approve or reject applications for large livestock operations but could not impose on them conditions of their own making, aside from shelterbelts or manure storage covers. They could not even impose conditions, already common, requiring farmers to inject manure into the soil rather than spread it on the surface of their fields.

The government justified such restrictions on municipal freedom on the grounds that local control must be balanced against provincial interests. In the words of Intergovernmental Affairs Minister Scott Smith, 'there are certain overriding principles that are critically important to the government.'[74] One such overriding principle, apparently, was that livestock farmers required a predictable planning process. Restricting municipal freedom, the minister's press secretary explained, would provide greater certainty – it would let farmers know what they were getting into.[75] The point was debatable. Allowing municipalities and farmers to negotiate mutually acceptable odour controls might have provided more certainty to farmers than forcing municipalities to reject unacceptable proposals outright. But the province may have been confident that, when faced with a take-it-or-leave-it choice regarding a proposed farm, a municipality would take it. It may well have been banking on the theory that a municipality would find it politically difficult to reject an application that conformed with its development plan – a plan that, being abstract, would have attracted less scrutiny and fewer objections than a specific application affecting identifiable parties.

Bill 33 met with only muted criticism from the parties in opposition, with the Liberals proposing amendments that would make the review process more open and enable municipalities to require manure injection, and the Progressive Conservatives proposing amendments that would streamline the review process, give farmers more certainty, and expand farmers' influence on the Technical Review Committees.

By far the most passionate criticisms of Bill 33 came from concerned citizens. At the Standing Committee hearings into the bill, many objections focused on the process created by it. Several presenters criticized the lack of public input into the review process. One economist charged, 'for rural Manitobans, Bill 33 represents another cobblestone in the "Road to Serfdom."'[76] Somewhat less dramatically, another presenter complained, 'Clause by clause, this bill chips away at rural democracy.' She challenged the legislators, 'How dare you presume to know what is best for our communities?'[77] Other objections focused on the probable outcome of the review process, with one presenter posing the question, 'Who are we planning for?'[78] Another suggested an answer: 'Changes in land use planning related to agriculture have one intent, to smooth the way for factory hog farms to locate in rural municipalities.'[79]

The agricultural community certainly seemed confident that the bill would smooth its way. Manitoba Pork Council – frustrated by years of working within a system that it described as confusing, duplicative, fractious, and divisive – called the bill 'long overdue.'[80] Manitoba Cattle Producers – likewise concerned about municipalities' arbitrary, subjective, or ideological decisions – praised the government's efforts to bring order to the planning process.[81] Dairy farmers and egg producers also voiced their support for the bill.

In all, the Standing Committee heard eighteen presentations and received eight written submissions on Bill 33. Not one weakened its resolve to change the planning regime. The bill was reported without amendment; it received Royal Assent on 16 June 2005. Thus ended

Manitoba's 33-year campaign to centralize the regulation of agricultural pollution. The transition to provincial control began in 1972, with the amendments to the Clean Environment Act authorizing the minister to override orders issued by the province's Clean Environment Commission, and continued with the enactment of the 1976 Nuisance Act, the country's first right-to-farm act. With the passage of the Planning Act, the transformation was complete.

7 Reversing the Trend: Decentralizing the Regulation of Agricultural Pollution

With the right to own, manage, and use natural resources comes the duty to prevent environmental harm and to protect the rights of people.
                                                                    – The Earth Charter

You've got the right to farm, but when the smell and runoff come on to my land, your rights end at your property line. If you can contain it there, fine. If not, it's my business.

                                                        – Organic farmer Hugh Doyle

In the last three decades, all of Canada's provinces have adopted some form of right-to-farm legislation. Manitoba led the way with its 1976 Nuisance Act. Quebec followed two years later with An Act to Preserve Agricultural Land. Newfoundland and Labrador was the last province to protect polluting farmers, proclaiming its Farm Practices Protection Act in 2003. Designed to sustain farmers and preserve farm land, all provincial right-to-farm laws shield farmers from legal liability for some of the environmental impacts of their operations. Manitoba's early legislation protected farms and other businesses from nuisance suits regarding odours. Quebec refined Manitoba's approach, protecting only pre-existing livestock farms that created odours or noise. New Brunswick added dust to its list of acceptable nuisances. Ontario eventually expanded its list to also include light, vibration, smoke, and flies.

Specific provisions of provincial right-to-farm laws vary widely. Some cover an entire province, while others apply only to designated agricultural regions. Some shield farmers from common-law nuisance actions, while others shield them from restrictive municipal by-laws as

well. Some protect all farmers, while others protect only those using
'normal' practices – effectively empowering the industry to set its own
standards[1] – or those complying with particular laws or regulations.
Some protect all farms, new and old, while others protect only established farms from the complaints of newcomers. Some instruct the
courts, limiting their discretion to rule against nuisances, while others
establish extrajudicial dispute resolution processes of varying levels of
independence and transparency.[2]

Although very different, the laws reflect remarkably similar thinking
about agriculture. Farmers have persuaded legislators in every party of
the overarching importance of their industry and the acceptability –
and inevitability – of its adverse environmental impacts. They have
convinced them that the common-law regime that regulated agriculture for centuries poses unacceptable threats to otherwise viable contemporary farms. That such thinking reflects self-interest rather than
reality has been lost on most provincial lawmakers, who seem oblivious to the enormous and unsustainable costs of contemporary agriculture – an industry that is not only one of Canada's major polluters but
also a significant drain on the public purse, requiring subsidies that far
exceed farmers' earnings.[3]

Right-to-farm legislation provides yet another subsidy to farmers by
allowing them to 'externalize' many environmental costs – to impose
costs on downwind or downstream individuals, communities, and
industries, rather than bearing them themselves. The right to farm, in
short, has become confused with a right to pollute. Farmers who
believe that they have acquired a right to pollute now demand compensation for measures taken to reduce pollution. The Ontario Federation
of Agriculture is typical in justifying its demand for assistance on the
grounds that 'if farmers are being called upon to make improvements
to their facilities in order to protect the environment, everyone in our
society will benefit, not just the farmers.'[4] The federation apparently
fails to appreciate the distinction between requiring farmers to provide
benefits and requiring them to stop producing harms. Just the former
warrants compensation. Nonetheless, programs that pay farmers for
so-called environmental services – including such harm-stopping measures as manure storage and run-off control – are increasingly common,
both federally and provincially.

Only in a topsy-turvy right-to-pollute world do farmers demand
money to cease harming their neighbours. Under the common law,
money flowed in the opposite direction: farmers paid damages to

neighbours they harmed.[5] The right to compensation, and in many cases to an injunction to prevent further harm, inhered in Canadians' property rights. In relieving farmers of the need to compensate those they have harmed and to cease their harmful activities, right-to-farm laws violate these age-old rights. Justice Robert Sharpe of the Ontario Court of Appeal expressed his reservations about this aspect of Ontario's law:

> This Act represents a significant limitation on the property rights of land-owners affected by the nuisance it protects. By protecting farming opera-tions from nuisance suits, affected property owners suffer a loss of amenities, and a corresponding loss of property value. Profit-making ven-tures, such as that of the appellants, are given the corresponding benefit of being able to carry on their nuisance creating activity without having to bear the full cost of their activities by compensating their affected neigh-bours. While the Act is motivated by a broader public purpose, it should not be overlooked that it has the effect of allowing farm operations, prac-tically, to appropriate property value without compensation.[6]

While laws permitting expropriation without compensation may violate ancient common-law principles and contemporary norms, they remain constitutional in Canada. The same cannot be said in the United States, where the Constitution prohibits the taking of private property for public use without just compensation. In 1999, the Iowa Supreme Court condemned that state's right-to-farm law as flagrantly unconsti-tutional, explaining, 'When all the varnish is removed, the challenged statutory scheme amounts to a commandeering of valuable property rights without compensating the owners, and sacrificing those rights for the economic advantage of a few. In short, it appropriates valuable private property interests and awards them to strangers.'[7]

Permitting farmers to commandeer others' property rights conflicts not only with deep-seated concepts of justice but also with environmen-tal sustainability. It violates the principle of polluter pay – a principle described by the Supreme Court of Canada not only as 'firmly entrenched in environmental law in Canada' but also as 'recognized at the international level.'[8] In insulating farmers from the costs of pollut-ing, right-to-farm laws remove incentives to operate sustainably. Farm-ers who can transfer to others the environmental, health, or social costs of their activities have few reasons to invest in technologies or choose management practices that reduce these costs. It is little wonder that three quarters of Canada's livestock farms use no method whatsoever to control odours from barns.[9] Although complaints about odours are common, and although farmers can pressure-wash feeding floors, adjust feed rations to reduce nitrogen and sulphur in manure, install filters on exhaust fans, create windbreaks, or take countless other measures to reduce odours, few have sufficient incentives to do so. Farmers likewise lack compelling reasons to minimize odours while storing or spreading manure. Covering manure storage facilities and injecting manure into the soil or incorporating it immediately after spreading can reduce odours by as much as 90 per cent.[10] But such practices can be costly, and many farmers, given the choice, would prefer to invest their funds elsewhere. Nor can most farmers justify installing composters, anaerobic digesters, aerators, or other manure treatment systems when 'normal' methods of handling manure – methods that involve no treatment at all – are deemed acceptable under right-to-farm regimes.

The perverse incentives created by right-to-farm laws may result in pollution that goes beyond permitted nuisances. In forgoing opportunities to reduce odours, farmers are often dispensing with practices that would benefit the greater environment, as well. Covering manure storage facilities, for example, would help prevent run-off from contaminating nearby waters. Composting manure would reduce the amount of methane – a potent greenhouse gas – released into the atmosphere. The right-to-pollute mentality fostered by right-to-farm legislation enforces the assumption that such measures are unnecessary. More insidiously, the twisted thinking behind many right-to-farm laws has begun to inform other environmental laws, as well. In Ontario, the equation of normal practices with acceptable practices – a concept introduced in the 1988 Farm Practices Protection Act – now appears in the Environmental Protection Act, the Nutrient Management Act, and the Provincial Policy Statement under the Planning Act.

Not only have right-to-farm laws failed to control pollution, they have also failed to bring peace to rural communities. Rather than resolving conflicts, they have merely, by restricting private litigation and, in many cases, municipal authority, limited rural residents' means of responding to conflicts.[11] Thus, while common-law nuisance actions against farmers are now rare, disputes between farmers and their neighbours remain commonplace. The number of cases brought before right-to-farm boards across Canada steadily increased throughout the 1990s,[12] as did complaints to other agencies. In 2002–3, after Alberta's new Agricultural Operation Practices Act went into effect, the Natural

Resources Conservation Board received 1,019 complaints. The follow-ing year, the number of complaints increased to 1,108, 741 of which con-cerned odours or other nuisances. The NRCB was unable to resolve more than 200 of the complaints.[13] In Ontario, after the passage of the Farm Practices Protection Act, complaints sky-rocketed, soon dwarfing those mediated by the Farm Pollution Advisory Committee in the 1970s and 1980s. Conflicts continue to pervade rural areas. Between 1998 and 2004, the agriculture ministry received an average of 675 com-plaints each year, with the number of complaints regarding odours from livestock operations increasing.[14]

Farmers often argue that these complaints are unfair, since they come from city-slicker newcomers unfamiliar with the realities of rural life. It is a myth that conflicts arise primarily from demographic changes. Many conflicts occur not between farmers and sprawling suburbanites but between farmers and other farmers. Indeed, approximately one-half of farm-related nuisance complaints are made by farmers.[15] While rural demographics have certainly changed in recent decades, so, too, have farms and farming practices. According to Ontario's agriculture minis-try, the type and size of a farming operation are more strongly linked to conflicts with neighbours than are the characteristics of the commu-nity.[16] Livestock producers tend to generate more conflicts than do veg-etable producers. And, although size is by no means the sole indicator of environmental sustainability,[17] as livestock operations increase in size and intensity, they tend to generate more manure, more noxious emis-sions, and more complaints – not only from newcomers but also from long-established rural residents.

Those farmers who do predate their neighbours often claim moral authority to create nuisances on the grounds that neighbours who have knowingly come to nuisances have no right to complain about them. Such arguments cannot withstand the test of the common law in Can-ada. Farmers who have polluted without penalty in the past have merely enjoyed free benefits, in that their neighbours have chosen not to enforce their own rights. The farmers have not, however, acquired any *right* to pollute or to continue polluting in the future. Thus, coming to a nuisance is rarely a defence under Canadian common law. Courts have generally held that a nuisance is a nuisance, regardless of which party was there first.[18]

Under the common law, nuisance-creating farmers have several options. The most straightforward of these is to modify their practices so as not to harm others. Alternatively, they may bargain with those

they affect, negotiating compromises acceptable to all parties.[19] By offer-
ing offsetting benefits, they may obtain permission to continue creating
some nuisances. When dispute resolution proves impossible, farmers
may try to purchase the lands surrounding their operations to ensure
that they harm no one but themselves. They may rent their newly
acquired lands to those who accept their nuisances, or they may re-sell
them with easements, covenants, or other restrictions attached to the
deeds. Alternatively, they may move their operations to less controversial locations, or, if all else fails, cease operations entirely. Such solutions,
arrived at freely and fairly between farmers and their neighbours, epit-
omize decentralized decision making. The process maximizes farmers'
freedom to choose solutions that are most effective and efficient on their
particular farms, while respecting the rights of those directly affected by
their operations.

The customary common law is thus often – but by no means always
– the best way to ensure that farmers' activities satisfy the needs and
preferences of all affected parties. Common-law property rights work
best when a polluting farm can be identified, when a limited number of
victims can be identified, and when the harm is substantial. When
many people suffer minor, cumulative damages from many small polluters, no one has an incentive to sue. Each suit would be costly and
ineffective. Such cases call for government-made laws and regulations.
The challenge lies in identifying the appropriate level of government
and the appropriate kind of law or regulation for any given situation.

Governance models that empower 'communities of interest' can
ensure that those bearing the economic, environmental, or social costs
of farms have a say in their establishment and operation. Political
boundaries do not always best define communities of interest. People
living in different municipalities may share an airshed or a watershed
with a polluting farm and have a legitimate interest in diminishing its
adverse impacts.[20] That said, municipalities are often themselves communities of interest and can serve as logical and valid regulators.
Municipalities often bear costs created by farms. Their infrastructure
costs may rise with increases in truck traffic to and from large farms,
their water treatment costs may rise as source waters become contaminated by manure or chemicals, their income from property taxes may
fall with decreases in assessments on properties near polluting farms,
they may be saddled with clean-up costs in the event of a manure spill
at a defunct or bankrupt farm, or desired development may be pre-
cluded by health and environmental concerns.[21]

In addition to representing their own direct interests, municipalities are expected to represent the interests of their citizens. Many argue that they do so more faithfully than do upper levels of government, being smaller, more accessible, less political, and more accountable.[22] Municipal councillors have the practical advantage of being close to the matters they are empowered to regulate. They are intimately familiar with their particular region, with unique local conditions and cultural norms, and with the quirky ways in which people use local resources. They can tailor specific solutions to specific problems. They have personal incentives to get the solutions right: they must live with the effects of their decisions on their health, their quality-of-life, and their property values. Sharing a community with others who are directly affected by their decisions, they also have powerful social incentives to act responsibly.[23]

As is true of any level of government, however, there is no guarantee that municipalities will protect individual interests. They may be susceptible to vocal minorities or special interests, and they may make decisions that seem arbitrary, reflect preferences rather than rights, or protect the majority at the expense of the minority. It is therefore essential to balance the political tradition of democratic decision making with the common-law tradition of individual property rights. Only by preserving – or restoring – such rights is it possible to guard against the tyranny of the majority.

Empowering municipalities and other communities of interest to set rules regarding the use of shared air or water while maintaining affected individuals' common-law rights to injunctions and damages will create a regulatory environment far more diverse than any designed by provincial or federal bureaucrats. Both municipal regulation and the common law allow for a variety of solutions, adapted to meet local and individual concerns. Some local regulations, private arrangements, and court decisions will, of course, be more successful than others. The failures will be less far-reaching than the failures of more centralized regulations. They will not be imposed on as many people, nor will they survive for as long. Communities and courts will become laboratories, testing and adjusting new approaches in a flexible and resilient regulatory process.[24]

The diversity inherent in decentralized regulation appeals to several farming organizations. The Christian Farmers Federation explains, 'Ontario's rural landscape varies hugely. Ontario is too diverse for a one-size-fits-all approach. The basic provincial rules need to be customized to suit the local landscape ... Rural vitality depends on communities consciously distinguishing themselves from the many others.'[25] The Canadian Cattlemen's Association likewise calls for regulatory approaches that provide for the tailoring of rules to specific regions.[26]

Support for decentralization – or individual and community empowerment – seems to be growing among a broad array of interest groups. Advocates fall on all points of the political spectrum. Some, such as Marilyn Churley, an Ontario parliamentarian who has complained about municipalities losing control over large farms in their areas, are committed leftists.[27] Ms Churley's concerns may reflect a general shift in the left, as described by Naomi Klein – a shift away from seeking 'centralized state solutions to solve almost every problem' and towards 'empowering people to make their own decisions.'[28] Other advocates of decentralization are centrists. Quebec's Liberals have argued that the province's ninety-six regional governments should be allowed to declare moratoriums on new pork production.[29] The Ontario Liberals' Clean Water Act envisions protecting drinking water sources through a locally driven watershed-based process. Still other decentralizers are aligned with the right and have long promoted property-rights-based approaches to pollution control. The Ontario Property and Environmental Rights Alliance has objected to 'blanket legislation and increased regulatory impingement on rural lands,' arguing instead that pollution from farms should be adjudicated by the courts.[30]

Environmentalists are joining the chorus in support of decentralization, albeit not often in the context of agricultural regulation. The University of Victoria's Polis Project argues for 'giving greater power and control to local peoples.' One of the root causes of ecological decline, it maintains, 'is the dislocation of power from communities.'[31] The Canadian Environmental Law Association has promoted strong community involvement, suggesting that 'municipalities need to be given the tools and the resources to protect water sources within their boundaries.'[32] The organization was 'thrilled' with the 2001 Supreme Court decision upholding a municipal by-law restricting the use of pesticides, calling it a victory for local democracy.[33] The Sierra Club and Great Lakes United have called for Great Lakes recovery efforts to be driven by state and local officials rather than by higher levels of government or bodies reporting to them. Empowering lower levels of government, they have argued, is vital 'to ensure the highest degree of accountability and public involvement.'[34]

Some environmentalists have applied such thinking to the regulation of agricultural pollution. One of the environmental movement's most passionate defenders of decentralization is the U.S. environmental lawyer Robert F. Kennedy Jr, who has charged Alberta's Natural Resources Conservation Board with making it too easy for unpopular hog operators to gain approval. 'This is the first thing they do when they go into a new area,' Mr Kennedy explained. 'They make sure local people cannot object ... This is one-stop shopping for polluters.'[35] He elaborated, 'The only way to control this industry is local control, and local control has now been removed from the municipalities in Alberta.'[36] Mr Kennedy recommended 'add[ing] a new ingredient of citizen involvement.'[37] Of course, citizen involvement can hardly be described as 'new,' given the prominent role long played by citizens under the common law.

The growing appreciation of individual and community decision making is consistent with the principle of subsidiarity, under which decisions are made as closely as possible to the people themselves – where feasible, by individuals; otherwise, as required, by local communities or higher levels of government. The principle gained prominence in twentieth-century Roman Catholic teachings. Pope Pius XI confirmed 'that most weighty principle' in 1931: 'Just as it is gravely wrong to take from individuals what they can accomplish by their own initiative and industry and give it to the community, so also it is an injustice and at the same time a grave evil and disturbance of right order to assign to a greater and higher association what lesser and subordinate organizations can do.'[38] The U.S. Catholic Bishops weighed in on the issue in 1986. 'The primary norm for determining the scope and limits of governmental intervention is the "principle of subsidiarity,"' they wrote. 'This principle states that, in order to protect basic justice, government should undertake only those initiatives which exceed the capacities of individuals or private groups acting independently. Government should not replace or destroy smaller communities and individual initiative.'[39] Five years later, Pope John Paul II likewise urged respect for the principle, explaining that 'needs are best understood and satisfied by people who are closest to them.'[40]

The principle of subsidiarity, appreciating broadly divergent needs and interests, respects the autonomy of individuals and local communities. It is thus profoundly decentralizing, providing 'a bottom-up vision of self-governance.'[41] As interpreted by one commentator, the principle holds that 'nothing should be done by a larger and more complex organization which can be done as well by a smaller and simpler organization. In other words, any activity which can be performed by a more decentralized entity should be.'[42]

The principle of subsidiarity has taken root in the secular world of international governance. The European Union embraced the principle to regulate the exercise of power, relying on it to determine when the Union should take action and when it should instead leave matters to its member states. The treaty establishing the European Community stated, 'In areas which do not fall within its exclusive competence, the Community shall take action, in accordance with the principle of subsidiarity, only if and in so far as the objectives of the proposed action cannot be sufficiently achieved by the Member States and can therefore, by reason of the scale or effects of the proposed action, be better achieved by the Community.'[43] The EU elaborated upon the principle as follows: 'The subsidiarity principle is intended to ensure that decisions are taken as closely as possible to the citizen and that constant checks are made as to whether action at Community level is justified in the light of the possibilities available at national, regional or local level.'[44]

The term subsidiarity is gradually making its way into Canadian legal discourse. The Supreme Court of Canada discussed the concept in its 2001 judgment confirming a municipality's right to restrict pesticide use. 'The case arises,' it wrote, 'in an era in which matters of governance are often examined through the lens of the principle of subsidiarity. This is the proposition that law-making and implementation are often best achieved at a level of government that is not only effective, but also closest to the citizens affected and thus most responsive to their needs, to local distinctiveness, and to population diversity.'[45]

Subsidiarity, as embodied in notions of decentralization and empowerment, increasingly informs Canadian discussions of agricultural regulation. A regulatory regime that respects the principle cannot accommodate right-to-farm laws that disempower individuals and concentrate authority at the provincial level. Decisions about the environmental impacts of farms will be taken as closely as possible to citizens only when such laws are repealed, returning to rural residents their common-law rights to enjoy their property without unreasonable interference and enabling them to defend these rights in court when necessary.

While such changes will restore citizens' access to the courts, it is by no means given that a rush of litigation against farmers will follow. The need to protect farmers from an onslaught of lawsuits has never been convincingly established. Nuisance actions were not common before the advent of right-to-farm legislation. Indeed, some provinces, such as Prince Edward Island and Newfoundland and Labrador, had never seen nuisance actions against farmers.[46] The common law strongly discouraged lawsuits over minor nuisances, which generally fell under the 'live and let live' rule. When cases did proceed, the courts never rode rough-shod over farmers' interests. They regularly took into account the character of the neighbourhoods in which the disputes arose. Commentators agree that it was not litigation itself but the *potential* for litigation that prompted many right-to-farm laws.[47]

It is impossible to know how dispute resolution or farming practices would have evolved had the common law been allowed to stand, or to predict how they will evolve when right-to-farm laws are repealed. One can reasonably assume some increase in litigation. After Iowa's Supreme Court declared the state's right-to-farm law unconstitutional, nuisance suits proliferated. The numbers, however, hardly became unmanageable. Suits against hog farms peaked in 2003, when thirteen were filed, and then declined to just two in 2004 and another two in the first nine months of 2005.[48] A few of the lawsuits have been dramatic. In a case decided in 2002, a jury awarded more than $33 million to four families neighbouring the state's largest hog operation – $1.06 million in actual damages, including pain, suffering, and reduced property values, and $32 million in punitive damages. An out-of-court settlement later eliminated the punitive damages.[49] Other damage awards have been more modest – and more in keeping with those that Canadian courts might issue.

Lawsuits will not be undertaken lightly. The time it takes for a case to make its way through the courts will deter plaintiffs who require quick solutions. The cost of a civil suit – and the risk of having to pay the defendant's costs if the suit is unsuccessful – will deter plaintiffs who have weak cases. Class-action suits, contingency fees, and the support of public-interest law firms or environmental organizations can help spread some of the costs and risks of litigation. But in the absence of tort reforms that make the court system still more accessible, affordable, and effective, lawsuits will generally be launched only as a last resort, after other attempts at dispute resolution have failed.

Furthermore, even when lawsuits are feasible, they will often be unnecessary. Common-law rights have a strong prophylactic effect. Once rights are clearly established, and are seen to be enforceable, farmers will adjust their practices in order to avoid interfering with those rights. Few will risk substantial investments in operations that may be subject to large damage awards or that may be shut down by those who are harmed by them. Recent lawsuits in Iowa have prompted the industry to rethink its creation of nuisances. Iowa Pork Producers Association advises farmers to take all reasonable steps – including keeping facilities clean, injecting or quickly incorporating manure, planting shelterbelts, and even purchasing neighbouring lands – to minimize the impacts of their operations on their neighbours. The association also notes that farmers may want to acquire nuisance insurance – the premiums for which would introduce further incentives to operate responsibly.[50]

Under a regime of renewed common-law liability, many aspects of farm operations will, of course, continue to require centralized regulation. Impacts that affect an entire watershed can often be managed at the watershed level. Impacts that are national in nature will often be best managed nationally. To take but one example, animal diseases may affect the larger agricultural community more severely than they do a farm's immediate neighbours. Cattle farmers across Canada suffered from the discovery, in 2003, that a cow on a family farm in Alberta was infected with BSE, or mad-cow disease. When the United States banned imports of Canadian cattle, prices plummeted, threatening to destroy the country's beef industry; the costs – to the industry and taxpayers – of one sick cow are thought to have exceeded $5 billion.[51] Western Canada's poultry industry was likewise threatened by the avian flu that spread rapidly through BC's Fraser Valley in April 2004, prompting the federal government to order the destruction of 19 million chickens, turkeys, geese, and ducks. The cull, affecting approximately 600 farmers, was expected to cost the industry $45 million.[52] Such impacts would, of course, be dwarfed should an avian flu virus mutate into a virulent human form. The director of the U.S. Centers for Disease Control has warned that avian flu poses the greatest health threat in the world.[53] It is appropriate for upper-level governments to regulate practices that can have such far-reaching impacts.[54]

The need for upper-level regulation, however, does not diminish the fact that individuals and communities, effectively empowered, will force farmers to curb practices that could otherwise have broad environmental impacts. The benefits of locally devised rules governing odours will extend well beyond their targets. Farmers who, in order to control odours, install biofilters on barns' ventilation systems, cover manure storage structures, and inject manure into the soil rather than spreading it on the surface will also, advertently or inadvertently, reduce emissions of harmful bioaerosols, capture ammonia and sulphur compounds that could contribute to smog or acid rain, curb greenhouse gas emissions, and prevent manure overflows during heavy rains, helping protect nearby waters from nitrogen and phosphorus pollution. Odour control thus becomes a proxy for other kinds of regulation, and local individuals and communities become proxies for all who are affected by agricultural pollution.

When it is necessary to impose provincial or federal laws and regulations, it will often be possible to incorporate into their design common-law principles that make them more equitable and effective. The most important common-law principle is expressed in the maxim, use your own property so as not to harm another's. This sweeping rule grants no farmer a right to pollute. Nor does it confuse normal and acceptable farming practices: pollution that harms others, regardless of whether it results from normal practices, is unacceptable. Statutes and regulations incorporating this principle will create virtual fences around farms. They will erect legal barriers against pollution, shielding neighbours from harms caused by farmers rather than protecting polluting farmers from their aggrieved neighbours, as do right-to-farm laws.

Statutes and regulations can also be crafted to reflect the common-law practice of considering specific circumstances – albeit while remaining governed by clear and predictable principles. In contrast to the broad brush generally preferred by legislators and bureaucrats, common-law courts have traditionally used finer instruments, examining the reasonableness of a disputed activity, the character of the neighbourhood in which it occurred, and, most important, the extent of the harm caused by it. They have emphasized not inputs, which are readily standardized, but outcomes, which vary widely. The heterogeneity of farms, farming practices, and their environmental impacts often demands such a targeted approach.[55] Replacing one-size-fits-all rules with those tailored to specific situations and focusing them on ends rather than means will, in effect, decentralize even centralized regulation.

Those designing statutes and regulations will also do well to incorporate the common law's approach to penalties for those who break the rules. Courts frequently award damages in successful common-law cases. Damages, like fines for non-compliance levied under many statutes, punish wrongdoers and deter unwanted behaviour. But unlike fines, which usually go to government coffers, damages go to those who have sued. They thus compensate individuals for the harms they have suffered. The common law also allows for injunctions – court orders requiring polluters to cease harmful activities or to take specific mitigating actions. Unlike fines, injunctions allow those who have been harmed to stop pollution and to prevent its recurrence. Because injunctions are enforced by those to whom they are awarded, they also allow for bargaining between parties and encourage compromises that suit all sides. Incorporating such remedies into provincial and federal statutes would further decentralize the regulation of agricultural pollution, empowering affected parties to negotiate locally appropriate solutions.

There are, of course, myriad approaches to decentralizing the regulation of agricultural pollution. Restoring genuinely decentralized decision making will require not only the repeal of right-to-farm laws but also the revision of laws addressing planning, nutrient management, drainage, and a number of other environmental issues – always with an eye towards empowering individuals and communities of interest to prevent or correct unsustainable practices. Such broad reforms will by no means be easy for the agricultural industry. Some polluting farmers will doubtless be put out of business; some low-value farmland will doubtless give way to higher-value uses. Truly viable farmers will succeed in internalizing the costs of their operations. They will create a new agricultural industry – an industry that respects the rights of others and is ecologically sustainable.

Notes

1. Canada's Farmers: Salt of the Earth or Assaulting the Earth?

The title of this chapter paraphrases organic farmer Tony McQuail, quoted by Brian McAndrew in 'Manure from factory farms causes concern,' *Toronto Star*, 6 June 2000.

1 Johanne Gélinas, Commissioner of Environment and Sustainable Development, 'You try living down here,' *Globe and Mail*, 3 October 2001.
2 Dennis O'Connor, *Report of the Walkerton Inquiry, Part One: The Events of May 2000 and Related Issues* (Toronto: Ontario Ministry of the Attorney General, January 2002), 13, 129. Cattle manure was just one of several causes of the Walkerton tragedy. Inept management of the afflicted water utility and inadequate oversight by provincial regulators also played key roles.
3 Alanna Mitchell et al., 'Fear of farming,' *Globe and Mail*, 3 June 2000.
4 G.L. Fairchild et al., 'Groundwater Quality,' in *The Health of Our Water: Toward Sustainable Agriculture in Canada*, ed. D.R. Coote and L.J. Gregorich (Ottawa: Agriculture and Agri-Food Canada Research Branch, 2000), Publication 2020/E, 71–3; and Bill Paton and Glen Koroluk, 'Brandon University scientist and environmentalists urge Manitoba government to protect groundwater drinking sources,' press release, 21 September 2005. One 1994 Alberta study reportedly found that 68 per cent of the wells tested were contaminated with bacteria. Al Mussell and Larry Martin, *Manure as a Public Health Issue* (Guelph: George Morris Centre, June 2000), 2, citing M.J. Goss et al., 'Contamination of Ontario Farmstead Domestic Wells and Its Association with Agriculture,' *Journal of Contaminant Hydrology* 32 (1998): 267–93.

5 The testing and treatment of well water can be expensive, and conventional treatment with chlorine can be ineffective in the face of significant contamination.

6 Parminder Raina et al., 'The Relationship between *E. coli* Indicator Bacteria in Well-water and Gastrointestinal Illness in Rural Families,' *Canadian Journal of Public Health* 90.3 (May–June 1999): 172–5.

7 Andrew Nikiforuk, 'Health Canada saw danger to district,' *Globe and Mail*, 31 May 2000.

8 Barbara Kermode-Scott, 'Air, Water Quality Need Closer Watch in Feedlot Alley,' *Medical Post* 37.3 (23 January 2001); and Andrew Nikiforuk and Danylo Hawaleshka, 'Should We Fear the Factory Farm?' *Reader's Digest* (June 2001).

9 Randy Christensen and Ben Parfitt, *Watered Down* (Vancouver: Sierra Legal Defence Fund, 2003), 27, 32, 38, 40, 43, 44.

10 International Joint Commission, *Twelfth Biennial Report on Great Lakes Water Quality,* September 2004, 24, http://www.ijc.org/php/publications/html/12br/english/report/index.html (accessed 9 November 2006).

11 Statistics Canada, 'Livestock, by Provinces,' 2001 Census of Agriculture, http://www40.statcan.ca/101/cst01/agrc32a.htm (accessed 9 November 2006).

12 Darrin Qualman and Fred Tait, *The Farm Crisis, Bigger Farms, and the Myths of 'Competition' and 'Efficiency'* (Ottawa: Canadian Centre for Policy Alternatives, 2004), 10, 26, 27.

13 In Ontario, for example, the contamination of rural wells with excessive levels of E. coli bacteria may have doubled between the 1950s and the 1990s. (Fairchild et al., 'Groundwater Quality,' 65, 72.)

14 Environment Canada, 'Hog farm pleads guilty to environmental charges,' press release, 17 February 2003.

15 Statistics Canada, *Cattle Statistics*, 2006, vol. 5, no. 1 (catalogue no. 23-012-XIE); and Statistics Canada, *Hog Statistics*, 2006, vol. 5, no. 1 (catalogue no. 23-010-XIE).

16 Statistics Canada, *A Geographical Profile of Manure Production in Canada, 2001*, January 2006 (catalogue no. 21-601-MIE – No. 077), 6, 15.

17 Commissioner of the Environment and Sustainable Development, 'Great Lakes and St. Lawrence River Basin,' *Report of the Commissioner of the Environment and Sustainable Development to the House of Commons* (Ottawa: Office of the Auditor General of Canada, October 2001), 145. The George Morris Centre points out that this comparison reflects the amount of solids in animal manure and human sewage rather than the volume of the waste. The volume of human waste is greatly increased by the water that is used to

transport and treat it. In terms of volume, the livestock in Ontario and Quebec produce waste equivalent to that of approximately 854,000 people. Al Mussell, Cher Brethour, and Larry Martin, *What the Environmental Commissioner Said: The Federal Report Card on Agriculture in Ontario and Quebec* (Guelph: George Morris Centre, November 2001), 2–3.

18 North Carolina State Representative Cindy Watson, Speech to joint annual meeting of Iowa Council and Soil and Water Conservation Society, 12 September 1998, http://www.iaenvironment.org/archivespdf/q_1998f_staggering_waste.pdf (accessed 9 November 2006).

19 Lucie Bourque and Robert Koroluk, 'Manure Storage in Canada,' in Statistics Canada, *Farm Environmental Management in Canada*, 2003, vol. 1, no. 1 (catalogue no. 21-021-MIE2003001).

20 Commissioner of the Environment and Sustainable Development, 'Great Lakes and St. Lawrence River Basin,' 149.

21 Amie Fulton et al., *It's Hitting the Fan* (Toronto: Environmental Defence Canada, October 2002), 20. In August 2005, a ruptured lagoon at a New York dairy farm spilled three million gallons of liquid manure into a drainage ditch and, from there, into the Black River.

22 Martin S. Beaulieu, 'Manure Management in Canada,' in Statistics Canada, *Farm Environmental Management in Canada*, 2004, vol. 1, no. 2 (catalogue no. 21-021-MIE – No. 002).

23 P.A. Chambers et al., 'Surface Water Quality,' in *The Health of our Water: Toward Sustainable Agriculture in Canada*, ed. D.R. Coote and L.J. Gregorich (Ottawa: Agriculture and Agri-Food Canada Research Branch, 2000), Publication 2020/E, 50–2.

24 Commissioner of the Environment and Sustainable Development, 'Great Lakes and St. Lawrence River Basin,' 145, 156. According to the commissioner (145, 171), 71 per cent of the farmland in Quebec and Ontario 'had much higher nitrogen levels in 1996 than in 1981. On more than 30 per cent of farmland, the levels of residual nitrogen pose a risk of water contamination.' In 2005, the commissioner reported that 'between 1981 and 2001, overall nitrate concentrations in water bodies in Canada increased by 24 per cent.' Commissioner of the Environment and Sustainable Development, 'Environmental Petitions,' in *Report of the Commissioner of the Environment and Sustainable Development to the House of Commons* (Ottawa: Office of the Auditor General of Canada, September 2005), 22.

25 See Robert Ferguson, 'Saving Canada's sickest lake,' *Globe and Mail*, 11 December 2004.

26 Health Canada, 'Well-water Contaminants: Results from PEAS,' in *Farm Family Health*, fall 1997.

27 Fairchild et al., 'Groundwater Quality,' 73. Nitrates, primarily from inten-
    sive poultry operations and raspberry production, have contaminated one
    half of the Abbotsford-Sumas Aquifer, which supplies drinking water to
    100,000 people in BC's Fraser Valley and in the United States. A 1992 sur-
    vey of almost 1,300 Ontario farm wells found that nitrate levels in 14 per
    cent exceeded drinking water guidelines (63, 65).
28 G. van der Kamp and G. Grove, 'Well Water Quality in Canada,' in *Proceed-
    ings of the 54th Canadian Geotechnical Conference and 2nd Joint IAH-CNC and
    CGS Groundwater Specialty Conference*, ed. M. Mahmoud et al. (Calgary,
    2001). A small survey of wells in Quebec's potato-growing regions found
    that almost 64 per cent exceeded guidelines for nitrates. Indicators and
    Assessment Office of Environment Canada, 'Nutrients in the Canadian
    Environment,' in *Reporting on the State of Canada's Environment*, undated
    (2001 or later), Table 4. Data for the 565 manure storage facilities in Mani-
    toba that are monitored by groundwater monitoring systems indicated that
    57 per cent of the facilities had contaminated groundwater with nitrate,
    chloride, ammonium, or sodium (Paton and Koroluk, 'Brandon Univer-
    sity.')
29 Graham Harrop, 'Backbench,' *Globe and Mail*, 7 September 2002.
30 It is thought that odour results from more than 165 compounds and their
    interactions. Alberta Agriculture, Food and Rural Development, *Dealing
    with Livestock Odour Concerns*, undated factsheet, http://www1.agric.
    gov.ab.ca/$department/deptdocs.nsf/all/epw8447/$file/AFRDOdour
    Factsheet1.pdf?OpenElement (accessed 23 November 2006).
31 Fulton et al., *It's Hitting the Fan*, 24, citing Hog Environmental Management
    Strategy Steering Committee, Situation Analysis, ManureNet (London:
    Agriculture and Agri-Food Canada, 1998).
32 Susan Schiffman et al., 'The Effect of Environmental Odors Emanating
    From Commercial Swine Operations on the Mood of Nearby Residents,'
    Brain Research Bulletin 37.4 (1995): 369.
33 A. Dennis McBride (North Carolina State Health Director), 'The Associa-
    tion of Health Effects with Exposure to Odors from Hog Farm Operations,'
    Medical Evaluation and Risk Assessment (Raleigh: North Carolina Depart-
    ment of Health and Human Services, 1998); Susan Schiffman et al., 'Poten-
    tial Health Effects of Odor from Animal Operations, Wastewater Treatment,
    and Recycling of Byproducts,' *Journal of Agromedicine* 7.1 (2000): 9; and Iowa
    State University and The University of Iowa Study Group, *Iowa Concen-
    trated Animal Feeding Operations Air Quality Study: Final Report*, February
    2002, 130, http://www.public-health.uiowa.edu/ehsrc/CAFOstudy/
    CAFO_final2-14.pdf (accessed 9 November 2006).

34 Susan Schiffman et al., 'Health Effects of Aerial Emissions from Animal
    Production Waste Management Systems,' in *Addressing Animal Production
    and Environmental Issues: Proceedings of International Symposium*, ed. G.B.
    Havenstein (Raleigh: North Carolina State University, 2001), 103–13.

35 Iowa State University and The University of Iowa Study Group, *Iowa Con-
    centrated Animal Feeding*, 6, 122–36; and Dana Cole et al., 'Concentrated
    Swine Feeding Operations and Public Health: A Review of Occupational
    and Community Health Effects,' *Environmental Health Perspectives* 108.8
    (August 2000): 688–91.

36 K. Thu, K. Donham, et al., 'A Control Study of the Physical and Mental
    Health of Residents Living Near a Large-scale Swine Operation, *Journal of
    Agricultural Safety and Health*, 3.1 (1997): 13–26; Steve Wing and Susanne
    Wolf, 'Intensive Livestock Operations, Health, and Quality of Life among
    Eastern North Carolina Residents,' *Environmental Health Perspectives* 108.3
    (March 2000): 233; Iowa State University and The University of Iowa Study
    Group, *Iowa Concentrated Animal Feeding*, 6–7, 137–8; and University of
    Iowa Department of Internal Medicine, 'UI Study: CAFOs near schools
    may pose asthma risk,' news release, 21 June 2006.

37 Kendall Thu, *Neighbor Health and Large-scale Swine Production*, undated
    [2000 or later], http://www.cdc.gov/nasd/docs/d001701-d0011800/
    d001764/d001764.pdf (accessed 9 November 2006).

38 Little is known about the distances travelled by harmful emissions. One
    study confirmed that bacteria from a henhouse travelled 200–300 metres
    downwind; another found elevated levels of particles and endotoxins 115
    metres downwind of a piggery. J. Seedorf and J. Hartung, *Emission of Air-
    borne Particulates from Animal Production*, Workshop on Sustainable Ani-
    mal Production, Germany, 2002, http://www.agriculture.de/acms1/
    conf6/ws4dust.htm (accessed 9 November 2006). A study conducted in
    2001 by the University of Saskatchewan found that endotoxins and air-
    borne microbial DNA from a swine barn housing 600 animals dissipated
    fairly quickly upon leaving the facility; these pulmonary irritants became
    diluted to levels associated with fresh air 600 metres downwind from the
    barn. (Veterinary Infectious Disease Organization, 'Swine barns not a
    downwind health threat,' press release, 6 March 2002.) In contrast, a pro-
    fessor of environmental health in Iowa reports that adsorbed malodorous
    vapours and gases appear to travel up to a kilometre from their source.
    Peter Thorne, 'Environmental Health Impacts of Concentrated Feeding
    Operations: Anticipating Hazards – Searching for Solutions,' *Environmen-
    tal Health Perspectives Online*, 14 November 2006, 7. Another study found
    that people within a two-mile radius of a 4000-sow facility reported sig-

nificantly higher rates of respiratory illness than did other rural residents. Thu, *Neighbor Health*.

39 Ibid.

40 Canadian Medical Association, '2002 General Council Resolutions,' http://www.cma.ca/index.cfm/ci_id/19816/la_id/1.htm (accessed 10 November 2006); and American Public Health Association, '2003 Policy Statements: 2003–7 Precautionary Moratorium and New Concentrated Animal Feed Operations,' *Association News*, no date, http://www.apha.org/legislative/policy/2003/2003-007.pdf (accessed 10 November 2006).

41 Michael Barza and Sherwood Gorbach, eds, 'The Need to Improve Antimicrobial Use in Agriculture: Ecological and Human Health Consequences,' *Clinical Infectious Diseases* 34, suppl. 3 (2002): S73; and P.A. Chambers et al., eds, *Linking Water Science to Policy: Effects of Agricultural Activities on Water Quality*, report on a workshop sponsored by the Canadian Council of Ministers of the Environment, Quebec City, 31 January and 1 February 2002, (Winnipeg: CCME, 2002), 12.

42 Cole et al., 'Concentrated Swine Feeding,' 688.

43 Iowa State University and The University of Iowa Study Group, *Iowa Concentrated Animal Feeding*, 11–12; and Cole et al., 'Concentrated Swine Feeding,' 692–3.

44 International Joint Commission, *Twelfth Biennial Report*, 30. Among the concerned experts are members of the Canadian Medical Association. In 1998, noting that 'addressing the issue of antimicrobial resistance has been called one of the most urgent priorities in the field of infectious disease,' and that 'the use of antimicrobial drugs in agriculture also has a significant impact on resistance in human pathogens,' the CMA called for a ban on the use of antibiotics for growth promotion in agriculture. Allison McGeer, 'Agricultural Antibiotics and Resistance in Human Pathogens: Villain or Scapegoat?' editorial, *Canadian Medical Association Journal*, 159.9 (1998): 1119–20.

45 Canadian Press, 'Restrict antibiotics in farming, panel urges,' *Globe and Mail*, 8 October 2002.

46 Maurice Korol, 'Fertilizer and Pesticide Management in Canada,' in Statistics Canada, *Farm Environmental Management in Canada*, 2004, vol. 1, no. 3 (catalogue no. 21-021-MWE2004002).

47 Chambers et al., 'Surface Water Quality,' 52–8, 60; and Fairchild et al., 'Groundwater Quality,' 67–70, 73. Of 212 Ontario farm ponds sampled in the 1970s and 1980s, 63 per cent were contaminated with pesticides. A 1994 survey of 103 farm dugouts in Alberta found herbicides in 48 per cent of them. A 1990 study detected herbicides in 38 per cent of the wells monitored in Nova Scotia. A 1992 Ontario survey detected herbicides in more

than 11 per cent of 1204 farm wells. Between 1999 and 2001, pesticides turned up in between 55 and 78 per cent of the samples collected from wells.in potato-growing areas of Quebec. Chambers et al., *Linking Water Science to Policy,* 8–9.

48 Jeff Wilson, 'P.E.I. needs to take another look at genetically modified potatoes,' *National Post,* 10 September 2002.

49 Canadian Environmental Law Association and Ontario College of Family Physicians (Environmental Health Committee), *Environmental Standard Setting and Children's Health,* August 2000, 8.

50 Natural Resources Defense Council, 'Pesticides Threaten Farm Children's Health,' 22 November 1998. Article based on Gina Solomon and Lawrie Mott, *Trouble on the Farm* (New York: NRDC, 1998).

51 Martin Mittelstaedt, 'Data point to breast-cancer risk,' *Globe and Mail,* 22 November 2002; 'Some pesticides may lead to Parkinson's disease,' *National Post,* 25 March 2003; Mayo Clinic, 'Study concludes that pesticide use increases risk of Parkinson's in men,' news release, 14 June 2006; Harvard School of Public Health, 'Pesticides exposure associated with Parkinson's disease,' news release, 26 June 2006; David Hopkins, 'Links between pollutants and cancers highlighted again,' *Edie,* 21 January 2005; and Marla Cone, 'Pesticides linked to low sperm count,' *Edmonton Journal,* 18 June 2003.

52 Canada's Environment Commissioner, acknowledging the absence of current estimates of the downstream costs of soil erosion in 2001, cited 1984 estimates of $91.2 million and added that recent information suggested that the downstream costs were higher than the costs to farmers – which exceeded $157 million in Ontario in 1986. Commissioner of the Environment and Sustainable Development, 'Great Lakes and St. Lawrence River Basin,' 157. According to one professor of environmental affairs, 'It is 600 times more expensive to remove suspended solids at treatments plants than it is to keep it [*sic*] on the farms in the first place.' Daniel Alesch, *New Strategies for Environmental Problems in Wisconsin: Breaking Out of the Box* (Thiensville, WI: Wisconsin Policy Research Institute, 1997), 2.

53 Agricultural water withdrawals, for irrigation and livestock watering, accounted for 9 per cent of Canadian water withdrawals in 1996. Environment Canada, *The Management of Water: Water Use, Agriculture,* July 2003. Nonetheless, as one water resource economist points out, agricultural water use is growing faster than other uses in Canada. Furthermore, the sector's water use is more consumptive than other sectors' – farmers return to the source less of the water they withdraw. Steven Renzetti, *Canadian Agricultural Water Use and Management,* Brock University Department of Economics Working Paper, March 2005, 4–5.

54 Environment Canada, 'Manure Causing White Haze,' *Science and Environment Bulletin* (May/June 1999): 7; also see Environment Canada, 'Study Examines Sources and Formation of Particulates,' *Science and Environment Bulletin* (November/December 2001): 3–4.

55 Claude Lagun, 'Greenhouse Gas Emissions and the Swine Industry' (Saskatoon: Prairie Swine Centre, 2004); Environment Canada, 'Nutrients in the Environment,' *Science and Environment Bulletin* (July/August 2001): 6–7; and Patricia Chambers et al., *Nutrients and Their Impact on the Canadian Environment* (Ottawa: Agriculture and Agri-Food Canada, 2001), vi.

56 Chambers et al., *Linking Water Science to Policy*, vii, 21; Chambers et al., *Nutrients and Their Impact on the Canadian Environment*, vi; and Environmental Commissioner of Ontario, *Thinking Beyond the Near and Now: 2002/2003 Annual Report* (Toronto: ECO, 2003), 40, 48.

57 Jennifer Lewington, 'Ministers to explore feedlot impact on water quality,' *Globe and Mail*, 5 June 2000.

58 Judith Lavoie, 'Federal NDP says Ottawa must assume responsibility for B.C.'s salmon farms,' *Victoria Times Colonist*, 26 June 2003.

59 Environmental Defence Canada has called on the federal government to establish minimum separation distances protecting residents and waterbodies; require manure management plans; establish standards for manure storage; ban non-medicinal uses of veterinary drugs and hormones; and subject intensive farms to the federal pollution registry. Fulton et al., *It's Hitting the Fan*, 3.

60 Leah Janzen, 'Hog farms take beating in documents,' *Winnipeg Free Press*, 21 March 2002.

61 O'Connor, *Report of the Walkerton Inquiry*, 131.

62 Gord Miller (Environmental Commissioner of Ontario), *The Protection of Ontario's Groundwater and Intensive Farming*, Special Report to the Legislative Assembly of Ontario (Toronto: ECO, 2000), 11.

63 Eva Ligeti (Environmental Commissioner of Ontario), *Open Doors: Ontario's Environmental Bill of Rights, 1998 Annual Report* (Toronto: ECO, 1999), 141.

64 Wayne J. Caldwell and Michael Toombs, *Planning and Intensive Livestock Facilities: Canadian Approaches*, paper presented to annual conference of Canadian Institute of Planners, June 1999, revised July 2000, 8.

65 John McCredie, Presentation to Ontario's Standing Committee on Resources Development, *Hansard*, 19 February 1998, 15:50.

66 Lisa Marr, 'Federal initiatives push for farm plans,' *Hamilton Spectator*, 25 June 2002.

67 Commissioner of the Environment and Sustainable Development, 'Great Lakes and St. Lawrence River Basin,' 174.

68 Commissioner of the Environment and Sustainable Development, 'Environmental Petitions,' 17, 20. The commissioner criticized Environment Canada for lacking a complete picture of whom it regulates, failing to gather data on a national basis in order to direct resources to the highest priority issues, and monitoring the impacts of its efforts. Agriculture Canada, she complained, has not developed its promised environmental management strategy and has failed to communicate its proposals and monitor its programs promoting beneficial management practices (18–26).

69 Jeremy Ashley, 'Water concerns aired at meeting,' *Belleville Intelligencer*, 20 January 2003.

2. Severing the Gold from the Dross: Using Common Law

1 *Malcolm Forbes Groat and Walter S. Groat v. The Mayor, Aldermen and Burgesses, being the Corporation of the City of Edmonton*, [1928] S.C.R. 522 at 532.

2 Henry of Bracton, cited by T.E. Lauer, 'The Common Law Background of the Riparian Doctrine,' *Missouri Law Review* 28.1 (winter 1963): 68.

3 *William Alfred's Case* (1611), 9 Coke Rep. f. 59a.

4 *Tenant v. Goldwin* (1703), 2 Ld. Raym. 1090, 92 E.R. 222 at 224 (K.B.).

5 Sir William Blackstone, *Commentaries on the Laws of England (1765–1769): Adapted to the Present State of the Law by Robert Malcolm Kerr* (London: John Murray, 1876), 3: 191.

6 *William Drysdale v. C.A. Dugas* (1896), 26 S.C.R. 20 at 23.

7 *Atwell et al. v. Knights* (1967), 61 D.L.R. (2d) 108 at 112, citing *Salmond on Torts*, 14th ed., 85.

8 *Star v. Rookesby* (1711), 1 Salk. 335, 91 E.R. 295 (K.B.).

9 Blackstone, *Commentories*, 3:185.

10 For a detailed discussion of trespass, nuisance, and riparian rights, see Elizabeth Brubaker, *Property Rights in the Defence of Nature* (London: Earthscan, 1995), chapters 1–3.

11 Blackstone, *Commentaries*, 3:186, 188.

12 *Ibid.*, 3:190–1.

13 *Weston Paper Co. v. Pope et al.*, 155 Ind. 394, 57 N.E. 719 at 721, 56 L.R.A. 899 (1900).

14 It is often difficult to determine whether the damages set by a court are adequate to compensate a plaintiff for his injury. Many injuries are not, to use one standard set out in *Shelfer v. London* (see below), 'capable of being estimated in money.' The costs arising from them may be individual, subjective, and changeable. A plaintiff can know far better than a court the value that he places on peace and quiet, or the amount of money he

would be willing to accept for breathing foul air. An injunction allows the plaintiff to negotiate with the defendant and to set his own price. The defendant may buy out the plaintiff (at a price acceptable to both parties); the parties may reach an agreement regarding mitigation (supplemented, perhaps, by compensation); or the defendant may simply exercise his right to the injunction.

15 *Shelfer v. City of London Electric Lighting Company and Meux's Brewery Company v. City of London Electric Lighting Company,* [1895] 1 Ch. 287 at 315–16.

16 Ibid., at 322.

17 *Duchman v. Oakland Dairy Co. Ltd.* (1928) Ontario Law Reports Vol. 63, 111 at 118, 119, 134.

18 *Atwell v. Knights,* at 111–12, citing *Salmond on Torts,* 14th ed., 84–5.

19 Ibid., at 113, citing *Shelfer v. London,* at 322.

20 *O'Regan and O'Regan v. Bresson, O'Regan and O'Regan* (1977), 23 N.S.R. (2d) and 32 A.P.R. 587 at 595–6, citing *Clerk and Lindsell on Torts,* 13th ed., 784–5.

21 Ibid., at 597.

22 *Breau, Morin, Lebel, Lebel, Lizotte, Bouchard, Carrier, Michaud and Paillard v. Soucy and Cyr (no. 1)* (1982), 41 N.B.R. (2d) and 107 A.P.R. 8. Translation of original French provided in 1983 appeal of same name, 52 N.B.R. (2d) and 137 A.P.R. 44 at 46.

23 Ibid., at 14, citing Fleming, *Law of Torts,* 4th ed., 1971, 346.

24 *Adams v. Provincial Grain Commission* (1990) 97 N.S.R. (2d) and 258 A.P.R. 411 at 425, citing *Pugliese et al. v. National Capital Commission* (1977), 17 O.R.(2d) 129 at 154.

25 *Newman v. Conair* (1972), [1973] 1 W.W.R. 316, citing *Crump v. Lambert* (1867), L.R. 3 Eq. 409 at 412.

26 *Metson v. R. W. DeWolfe Limited* (1980), 10 CELR 109 at 111, citing *Rylands v. Fletcher,* [1868] 3 L.R. 330 at 340.

27 Ibid., at 113. Well-water contamination from a racehorse training track was the subject of a 1990 British Columbia case. Water carrying leachate from the woodchip ground cover and coliform bacteria from horse manure and urine had seeped onto neighbours' property, polluting their well and making them ill. The judge applied the by now familiar formula: 'The water seepage and its effects clearly interfered with the plaintiffs' enjoyment of their land. I find that the interference was substantial and unreasonable. Thus, in law, the interference constitutes a nuisance.' *Vidler v. Page,* British Columbia Supreme Court, Vancouver Registry No. C892547, (1990) B.C.J. No. 2208.

28 *McClean v. Springborn,* British Columbia Supreme Court, Vernon Registry No. 8700246, [1988] B.C.J. No. 2172.

29 Edward Coke, *The First Part of the Institutes of the Laws of England* (London, 1628).

30 *Larry Hoffman and Dale Beaudoin v. Monsanto Canada Inc. and Aventis Cropscience Canada Holding Inc.*, Statement of Claim, Court of Queen's Bench, Saskatoon, January 10, 2002, at 7. In May 2005, a lower court judge denied the plaintiffs' application for certification of the lawsuit as a class action. In August 2005, Saskatchewan's Court of Appeal granted the plaintiffs leave to appeal the May ruling. Although an appeal was planned, no appeal date had been set at the time of this writing (May 2006).

31 For a discussion of how Canadian courts have, in the last century, strayed from a strict application of customary common-law principles and accepted previously unacceptable defences, see Brubaker, *Property Rights*, chapter 7.

32 *William Alfred's Case*, f. 58a.

33 *Breau v. Soucy*, at 15, citing *St. Helen's Smelting Co. v. Tipping* (1865), 11 H.L.C. 642, (1865) 11 E.R. 1483; and *Newman v. Conair*, at 321.

34 In choosing negligence over strict liability, a court would implicitly rule that the *effects* of a farmer's action were less important than his *intent*. In his discussion of the shortcomings of negligence theory (or the theory of 'reasonable conduct'), Murray Rothbard points out that the standard is vague and subjective and that it tends to favour the defendant over the plaintiff. Murray Rothbard, 'Law, Property Rights, and Air Pollution,' *Cato Journal* 2.1 (spring 1982): 64–7.

35 *Atwell v. Knights*, at 110 citing *Salmond on Torts*, 14th ed., 96.

36 *Breau v. Soucy*, at 15–16, citing *Adams v. Ursell* (1913), 1 Ch. 269 and *Pwllback Colliery Co. Ltd. Woodman* (1915), A.C. 634 at 638. In another dispute concerning a piggery – this one in Nova Scotia – the trial judge likewise recognized that the farmer could not defend himself on the grounds that he was doing everything possible to reduce odours: 'Unfortunately for the defendant his care in keeping the odour to a minimum does not necessarily mean that he is not liable for it.' He noted the emphasis, in the law of nuisance, on the kind of harm caused rather than on the kind of conduct causing it. Original decision cited by Appeal Division in *Fogarty and Crouse v. Daurie*, [1986] 77 N.S.R. [2d] and 191 A.P.R. 34 at 35.

37 *Metson v. DeWolfe*, at 111, citing *Rylands v. Fletcher*, [1868] 3 L.R. 330 at 340.

38 Ibid., at 113.

39 The exception to this under Canadian common law is when a farmer acquires a 'prescriptive right' – a legal right to carry on a long-standing activity. If a farmer has been creating a nuisance for twenty years, and no one has objected, he may have acquired a prescriptive right to create that

nuisance. A farmer whose operations have changed over the years could not claim such a right.

In the United States, a farmer who has been creating odours for far fewer than twenty years may have a right to continue to create odours, and a new neighbour may have no right to complain. The fact that the new neighbour 'came to the nuisance' will rarely be an absolute defence, but it may be one factor to be considered by the court. Regardless, the coming to the nuisance doctrine now influences U.S. courts far less than it has in the past. Timothy Swanson and Andreas Kontoleon, *2100: Nuisance*, 1999, 391, http://encyclo. findlaw.com/2100book.pdf (accessed 13 November 2006); and Dean Lueck, 'First Possession,' in *The New Palgrave Dictionary of Economics and the Law*, 1998. U.S. right-to-farm laws commonly codify the common-law defence of coming to a nuisance.

40 *Fleming v. Hislop* (1886), 11 App. Cas. 686 at 696-7.

41 The judge backed up his reasoning with a citation from a book on torts: 'The fact that the plaintiff has come to the nuisance does not prevent him from recovering damages.' *O'Regan v. Bresson*, at 590, citing *Clerk and Lindsell on Torts*, 13th ed., para. 1436, 810.

42 The distinction between a right and a benefit may be easier to grasp in the context of amenities than it is in the context of pollution. A landowner often enjoys an amenity – such as walking his dog on vacant land or having a beautiful view – without acquiring any right to it. He risks losing the amenity at any time; a neighbouring landowner may build on the land, precluding further dog walks or ruining the view. A landowner who wishes to maintain an amenity that he has previously enjoyed for free must secure a *right* to it – either by purchasing the land or by reaching an agreement with the owner of the land.

43 For further discussion of the ways in which the coming to the nuisance doctrine may encourage inefficient investment in a race to establish prior rights, see Swanson and Kontoleon, *Nuisance*, 392; Lueck, *First Possession*.; Donald Wittman, 'Coming to the Nuisance,' in *The New Palgrave Dictionary of Economics and the Law* (n.p.: MacMillan, 1998); and Rohan Pitchford and Christopher Snyder, 'Coming to the Nuisance: An Economic Analysis From an Incomplete Contracts Perspective,' Stanford/Yale Junior Faculty Forum Research Paper 01-17, February 2001, 2, http://papers.ssrn.com/sol3/ papers.cfm?abstract_id+280842 (accessed 13 November 2006).

44 The traditional rule that courts should not take the economic implications of their decisions into account was challenged by Nobel economist Ronald Coase in his famous article, 'The Problem of Social Cost.' Coase saw harm not as something inflicted by one party upon another but as reciprocal. In a

dispute, each party damages the other, or imposes unwanted costs on the other. Avoiding harming one party can inflict harm on the other. 'The real question that has to be decided is: should A be allowed to harm B or should B be allowed to harm A? The problem is to avoid the more serious harm.' In deciding who should be allowed to harm whom, Coase argued, courts should consider which party can remedy the problem at the lowest cost. They should seek the most efficient result – the assignment of rights that maximizes social value. Ronald Coase, 'The Problem of Social Cost,' *Journal of Law and Economics* 3 (October 1960): 2, 19–22.

Coase seemingly overlooked the impossibility of courts making accurate judgments about competing values. Value is often subjective, non-comparable, and incapable of being measured by an outsider. Accordingly, a court can maximize monetary or social value no more successfully than can any central planner. (Roy Cordato, *Chasing Phantoms in a Hollow Defense of Coase*, January 2000, 5–8, http://mises.org/journals/scholar/Cordato3.pdf (accessed 13 November 2006); and Walter Block, 'Coase and Demsetz on Private Property Rights,' *Journal of Libertarian Studies* 1.2, (1977): 111–15.

45 *Breau, Morin, Lebel, Lebel, Lizotte, Bouchard, Carrier, Michaud and Paillard v. Soucy and Cyr* (No. 128/82/CA) (1983), 52 N.B.R. (2d) and 137 A.P.R. 44 at 58. This issue also arose in the Nova Scotia piggery case. The judge hearing the appeal found that the trial judge was wrong to consider the burden that might be created by a finding of nuisance. The trial judge had written, 'To find for the plaintiffs on the evidence adduced before me would create an unbearable burden for the defendant and indeed for the entire farming community of Nova Scotia.' The appeal judge dismissed this as 'an incorrect consideration of the applicable burden of proof.' Having determined that the trial judge erred in applying the proper principles, the appeal judge ordered a new trial. *Fogarty and Crouse v. Daurie* (1986), 77 N.S.R. (2d) and 191 A.P.R. 34 at 37.

46 *O'Regan v. Bresson*, at 596, citing *Clerk and Lindsell on Torts*, 13th ed., 784–5.

47 Ibid., at 593, citing *Clerk and Lindsell on Torts*, 13th ed. paragraph 1396, 785.

48 *Adams v. Provincial Grain Commission* (1990), 97 N.S.R. (2d) and 258 A.P.R. 411 at 427, citing Fleming, *Law of Torts*, 3rd ed., 378.

49 The judge hearing the New Brunswick cottagers' complaint against local hog farmers likewise consulted a legal text on this issue. 'Legal intervention,' he found, 'is warranted only when an excessive use of property causes inconvenience beyond what other occupiers in the vicinity can be expected to bear, having regard to the prevailing standard of comfort of the time and place.' *Breau v. Soucy*, at 14, citing Fleming, *Law of Torts*, 4th ed., 346.

50 *O'Regan v. Bresson*, at 596, citing *Clerk and Lindsell on Torts*, 13th ed., 784–5.

51 *MacGregor v. Penner et al.* (1992), 82 Man.R. (2d) 178 at 180, citing *Royal Anne Hotel Co. v. Ashcroft*, [1979] 2 W.W.R. 462 (B.C.C.A.) at 466–8.

52 *Salmond on Torts*, 10th ed., 228–31.

53 *Adams v. Grain Commission*, at 426, citing *Kent v. Dominion Steel and Coal Corporation Ltd.* (1964), 49 D.L.R. (2d) 241 at 246.

54 *Miller v. Krawitz et al.*, W.W.R. [1931 Vol. 1], 577 at 581–2, citing *St. Helen's v. Tipping* and *Rushmer v. Polsue and Alfieri, Ltd.*, [1906] 1 Ch. 234, 75 L.J. Ch. 79.

55 *MacGregor v. Penner*, at 181–3.

56 *Smith v. Jaedel Enterprises (1978) Ltd.*, [1986] B.C.J. No. 1138.

57 *Clark v. Ward and Kirstein v. Ward* (1909), 9 W.L.R. 657 at 663, citing *Vaughan v. Menlove*, 3 Bing N. C. 468.

58 *Tubervil v. Stamp*, 1 Salk. 13, 12 Mod. 151, 1 Ld. Raym. 264, 1697, Holt King's Bench, UK.

59 *Dean v. McCarty*, 2 U.C.R. 448 at 450 (Queen's Bench, Hilary Term, 9 Vic.), 1846, Upper Canada Queen's Bench. It was rare for English or Canadian courts to put the public good before private rights. As Blackstone explained, 'So great ... is the regard of the law for private property that it will not authorize the least violation of it; no, not even for the general good of the whole community.' Blackstone, *Commentaries*, 1:109–10.

60 *Clark v. Ward*, at 660–1.

61 *Lickoch et al. v. Madu, etc.*, [1973] 2 W.W.R. 127 at 136.

62 Ibid., at 137, 132.

63 *Northwestern Utilities Ltd. et al. v. Standard Safety & Consulting Services Ltd. et al.* (1981), 35 A.R. 616 at 626–7. In 1976, the Alberta Supreme Court issued a similar decision – finding that a farmer's negligent burning of straw contributed to several accidents – in *Zaruk et al. v. and Schroderus et al.*, 71 D.L.R. (3d). The court determined that 'the ordinarily reasonable and prudent person' would have foreseen the danger of smoke and would have taken care to avoid creating it (at 224).

64 Lord Halsbury, *Canadian Pacific Ry. Co. v. Roy* (1901), C.R. [12] A.C. 374 at 389 (P.C.).

65 For a discussion of the evolution of the defence of statutory authority, and the erosion of common-law property rights, see Brubaker, *Property Rights*, chapters 4–7.

66 *Bennet v. Grand Truck R.W. Co. et al.* (1901), 2 O.L.R. 425.

67 Donald Dunn et al., *Ontario Right to Farm Advisory Committee Report*, submitted to the Honourable Jack Riddell, Minister of Agriculture and Food, 10 July 1986, 9.

68 For an extensive discussion of this issue, see the Supreme Court decision in
   Tock et al. v. St. John's Metropolitan Area Board (1989), 64 D.L.R. (4th) 620. In
   Tock, two judges questioned the wisdom of the inevitability test, and with it
   the value of the defence of statutory authority. Chief Justice Dickson con-
   curred with Mr Justice La Forest that inevitability itself should not excuse
   exemption from tort liability. The fact that an operation will inevitably
   damage some individuals does not explain why those individuals should
   be responsible for paying for that damage. 'Arguments about inevitability,'
   the judges agreed, 'are essentially arguments about money ... "Inevitable"
   damage is often nothing but a hidden cost of running a given system.'
   Their conclusion? 'The costs of damage that is an inevitable consequence of
   the provision of services that benefit the public at large should be borne
   equally by all those who profit from the service.' The judges added that
   requiring the body that provides a service to bear the costs of its operations
   could serve as a valuable deterrent: 'If the authority is to bear the costs of
   accidents ... it may realize that it is more cost-effective to forestal [*sic*] their
   occurrence.'

3. The Evolution of the Right to Farm in Manitoba

1 Legislative Assembly of Manitoba, *Hansard*, 12 June 1997, 11:40.
2 The background on their dispute is provided in *Lisoway v. Springfield Hog
   Ranch Ltd.*, [1975] M.J. No. 188, Manitoba Court of Queen's Bench, Wilson
   J., 24 November 1975.
3 Legislative Assembly of Manitoba, *Clean Environment Act*, sections 2 and
   17, assented to 20 July 1972.
4 Manitoba Regulation 34/73, Being a Regulation under the Clean Environment Act Respecting Livestock Production Operations, *Manitoba Gazette*,
   vol. 102, no. 8, 24 February 1973.
5 Lieutenant-Governor-in-Council of the Province of Manitoba, Order-in-
   Council No. 1107, 17 October 1973.
6 This remark was recalled in Legislative Assembly of Manitoba, *Hansard*,
   31 May 1976, 4458.
7 Clean Environment Commission, *Annual Report: 1973* (Winnipeg: CEC,
   1973), 8–9.
8 *Lisoway v. Springfield*, paragraph 6, citing *Salmond on Torts*, 16th ed., 51.
9 Ibid., paragraph 6, citing *Walter v. Selfe* (1851), 4 DeG. 6 Sm 315.
10 Ibid., paragraph 7, citing *Salmond on Torts*, 16th ed., 57.
11 Ibid., paragraph 8, citing *McKenzie v. Kayler* (1905), 15 Man. R. 660 at
   664.

12 Ibid., paragraph 8.
13 Ibid., paragraph 10, citing *Salmond on Torts*, 63.
14 Ibid., paragraph 11, with reference to *City of Portage la Prairie v. B.C. Pea Growers Ltd.*, [1966] S.C.R. 150.
15 Ibid., paragraph 34.
16 Ibid., paragraph 12, with reference to *City of Portage la Prairie v. B.C. Pea Growers Ltd.*, 156.
17 Ibid., paragraph 34, citing *McKie v. K.V.P. Company Limited*, [1948] O.R. 398, citing *City of Manchester v. Farnworth*, [1930] A.C. 171, at 203.
18 Ibid., paragraph 21.
19 Legislative Assembly of Manitoba, Bill 68, *The Nuisance Act*, section 2.
20 Markus Buchart, 'Who let the hogs out?' undated.
21 Legislative Assembly of Manitoba, *Hansard*, 31 May 1976, 4454–5.
22 Ibid., 4458–9.
23 Ibid., 4502.
24 Ibid., 4501.
25 At the time, the protections provided to Manitoba's farmers were the most comprehensive in North America. In 1963, Kansas passed a law protecting from nuisance suits feedlots that met specified standards. Oklahoma followed suit in 1969. A number of states passed broader right-to-farm laws in the late 1970s. All states now have some form of right-to-farm legislation. For information on the evolution of the right to farm in the United States, see Keith Burgess-Jackson, 'The Ethics and Economics of Right-to-Farm Statutes,' *Harvard Journal of Law and Public Policy* 9 (1986): 481–523; Adesoji O. Adelaja and Keith Friedman, 'Political Economy of Right-to-Farm,' *Journal of Agricultural and Applied Economics* 31.3 (December 1999): 565–79; Jacqueline P. Hand, 'Right-to-Farm Laws: Breaking New Ground in the Preservation of Farmland,' *University of Pittsburgh Law Review* 45 (1984): 289–350; and Neil Hamilton, 'Right to Farm Laws Reconsidered: Ten Reasons Why Legislative Efforts to Resolve Agricultural Nuisances May Be Ineffective,' *Drake Journal of Agricultural Law* 3.1 (spring 1998): 103–18.
26 Legislative Assembly of Manitoba, *Hansard*, 13 May 1992, 3325–6.
27 Ibid., 4186–7.
28 Ibid., 4891.
29 Ibid., 4893.
30 Legislative Assembly of Manitoba, *Hansard*, 15 May 1997, 10:10.
31 Jim Shapiro (President, St Germain-Vermette Community Association), presentation to the Standing Committee on Law Amendments, Legislative Assembly of Manitoba, *Hansard*, 17 June 1997, 19:20.

32 Legislative Assembly of Manitoba, *Hansard*, 15 May 1997, 10:10.
33 Ibid., 12 June 1997, 11:40.
34 Ibid., 26 June 1997, 17:30.
35 Ibid., 12 June 1997, 11:50.
36 Ed Tyrchniewicz et al., *Sustainable Livestock Development in Manitoba: Finding Common Ground*, chapter 2, 'Background,' report prepared for the Government of Manitoba by the Livestock Stewardship Panel, December 2000, 6, http://www.gov.mb.ca/agriculture/news/stewardship/livestock.pdf (accessed 13 November 2006).
37 Ibid., 7.
38 Lawrence Solomon and Carrie Elliott, *Agricultural Subsidies in Canada 1992–2001* (Toronto: Urban Renaissance Institute, 2002).
39 Tyrchniewicz et al., *Sustainable Livestock Development*, 4.
40 Ibid., 8.
41 Don Dewar (Keystone Agricultural Producers), presentation to Standing Committee on Municipal Affairs, *Hansard*, 21 June 2001, 18:50.
42 Ted Muir (Manitoba Pork Council), presentation to Standing Committee on Municipal Affairs, *Hansard*, 21 June 2001, 18:30–18:40.
43 Legislative Assembly of Manitoba, *Hansard*, 19 June 2001, 15:00.
44 Ibid. Emphasis added.
45 Ibid., 15:10.
46 Ibid.
47 The Farm Practices Protection Act, section 2, Acts, Regulations and By-laws, http://www.gov.mb.ca/agriculture/livestock/pork/swine/bah02s01.html (accessed 27 May 2004).
48 Remarks to the Standing Committee on Law Amendments, Legislative Assembly of Manitoba, *Hansard*, 17 June 1997, 19:40.
49 Legislative Assembly of Manitoba, *Hansard*, 19 June 2001, 15:10.
50 Leah Janzen, 'Expert warns of hog farm "disaster,"' *Winnipeg Free Press*, 6 October 2004. Between 1996 and 2002, hog production in the province more than doubled, increasing from 3.2 million to 6.65 million hogs. Tyrchniewicz et al., *Sustainable Livestock Development*, 8; and Manitoba Agriculture, Food and Rural Initiatives, *2002 Manitoba Agriculture Yearbook* (Winnipeg: MAFRI, 2004), 55.
51 Manitoba Agriculture, Food and Rural Initiatives, *Manitoba Swine Industry Facts*, October 2002, http://www.gov.mb.ca/agriculture/statistics/aac13s01.html (accessed 13 November 2006).
52 Tyrchniewicz et al., *Sustainable Livestock Development*, 10.
53 Ibid.
54 Ibid., 50.

4. The Legacy of Right-to-Farm Legislation in New Brunswick

1 Murray Corey, presentation to New Brunswick Select Committee on Agriculture and Renewable Resources, 5 December 1985; unedited transcript.
2 Chris Morris, Canadian Press, 'N.B. moves to protect farmers from nuisance neighbours,' *New Brunswick Telegraph Journal*, 8 March 2002.
3 CTV News Staff, 'Raising a stink,' *W-Five*, 15 November 2002.
4 Sarah Marchildon, 'Province hopes to lure farmers from Europe,' *New Brunswick Telegraph Journal*, 8 December 1998; and editorial, 'Hypocrisy stinks,' *New Brunswick Telegraph Journal*, 16 September 1999.
5 New Brunswick Department of Environment and Local Government, *Metz Farms 2 Ltd.: Surface Water and Groundwater Monitoring Results 2000–2001*, Environmental Reporting Series T2002–01. Italics added.
6 CBC, 'Metz solution,' nb.cbc.ca News, 8 March 2002.
7 Beyond Factory Farming, 'Air and water pollution evidence mounts against factory hog operation,' press release, 16 April 2004.
8 Communications New Brunswick, 'New Brunswick will provide assistance to Metz for technological solution,' news release, 8 March 2002.
9 Daniel McHardie, 'Agriculture minister kept hopping over Metz issue,' *Moncton Times and Transcript*, 16 March 2002.
10 Ibid.
11 CBC, 'Metz Farms launch high tech battle on smell,' nb.cbc.ca News, 12 July 2002.
12 Jean François Peltier (Department of Agriculture, Fisheries and Aquaculture), telephone conversation with Elizabeth Brubaker, 22 April 2004; and Cristelle Léger (Department of Environment), telephone conversation with Elizabeth Brubaker, 22 April 2004.
13 Rhonda Whittaker, 'Hog farm must treat stink,' *Times and Transcript*, 21 August 2004.
14 Kevin McKendy, telephone conversation with Elizabeth Brubaker, 4 October 2005.
15 Hans Kristensen, telephone conversation with Elizabeth Brubaker, 29 September 2005.
16 In May 2006, one local resident reported that it appeared that the manure lagoon was being emptied. However, the agriculture ministry declined to answer questions about the status of Metz 2. It refused to provide information about the emptying of the lagoon or the rehabilitation of the site. Nor would it confirm that the farm's licence had expired. Lynn Moore, telephone conversation with Elizabeth Brubaker, 29 May 2006.
17 Daniel McHardie, 'Lawsuit launched against 14 farms,' *Moncton Times and Transcript*, 17 January 2003.

18 *Desrosiers et al. v. Sullivan and Sullivan Farms Ltd.* (1986), 66 N.B.R. (2d) 243 (Q.B.), at 246.

19 Ibid., at 251. Also see Don McKay, 'Decision reserved in pig farm case,' *New Brunswick Telegraph Journal*, 9 October 1985.

20 *Desrosiers v. Sullivan*, at 249.

21 *Sullivan and Sullivan Farms Ltd. v. Desrosiers et al.* (1987), 76 N.B.R. (2d) 271 (C.A.), at 274-5. Italics added.

22 *Breau, Morin, Lebel, Lebel, Lizotte, Bouchard, Carrier, Michaud and Paillard v. Soucy and Cyr* (1982), 41 N.B.R. (2d) and 107 A.P.R. 8. Appeal of same name (1983) 52 N.B.R. (2d) and 137 A.P.R. 44.

23 New Brunswick Department of Agriculture and Rural Development, *A Discussion Paper on Right to Farm Legislation*, 1985, 6, 9, 17, 18, 20.

24 Tim Andrew, presentation to New Brunswick Select Committee on Agriculture and Renewable Resources, 3 December 1985 (all presentations in notes 24–33 and the statement in note 35 are from unedited transcripts); and David Meagher, 'Right-to-farm measures may be moved up a year,' *New Brunswick Telegraph Journal*, 4 December 1985.

25 Milton D'Aoust and Mr Rideout, presentation to New Brunswick Select Committee on Agriculture and Renewable Resources, 4 December 1985; and Canadian Press, 'Hog farmer plans to appeal court decision,' *New Brunswick Telegraph Journal*, 5 December 1985.

26 William Schrage, presentation to New Brunswick Select Committee on Agriculture and Renewable Resources, 4 December 1985.

27 Ralph Crossen, presentation to New Brunswick Select Committee on Agriculture and Renewable Resources, 5 December 1985.

28 George Slipp, presentation to New Brunswick Select Committee on Agriculture and Renewable Resources, 4 December 1985.

29 Bill Sherwood, presentation to New Brunswick Select Committee on Agriculture and Renewable Resources, 3 December 1985; and David Meagher, 'Legislation urged to protect province's farmland,' *New Brunswick Telegraph Journal*, 4 December 1985.

30 Warren Frazee, presentation to New Brunswick Select Committee on Agriculture and Renewable Resources, 3 December 1985; and David Meagher, 'He favors committee review of right-to-farm guidelines,' *New Brunswick Telegraph Journal*, 4 December 1985.

31 Marie Chambers, presentation to New Brunswick Select Committee on Agriculture and Renewable Resources, 5 December 1985.

32 Murray Corey, presentation to New Brunswick Select Committee on Agriculture and Renewable Resources, 5 December 1985.

33 Médard Bérubé, presentation to New Brunswick Select Committee on Agriculture and Renewable Resources, 30 January 1986; and Dan Toner, 'Farm-

ers, non-farmers square off,' *New Brunswick Telegraph Journal*, 31 January 1986.

34 Dan Toner, 'Neighboring residents: We were here first,' *New Brunswick Telegraph Journal*, 7 December 1985.

35 Mr Marmen, statement (as chairman) to New Brunswick Select Committee on Agriculture and Renewable Resources, 5 December 1985.

36 *Journal of Debates (Hansard) of the Legislative Assembly of the Province of New Brunswick*, session of 1986, vol. 9, 6 June 1986, 3455.

37 Ibid., 3455–7.

38 'Court ruling on farm results in resolution,' *New Brunswick Telegraph Journal*, 2 December 1985.

39 Mike Dillon (Land policy specialist with Department of Agriculture [retired]), telephone conversation with Elizabeth Brubaker, 2 June 2004.

40 Communications New Brunswick, 'New Brunswick government supports farming industry,' news release, 7 March 2002.

41 Lawrence Solomon and Carrie Elliott, *Agricultural Subsidies in Canada 1992–2001* (Toronto: Urban Renaissance Institute, June 2002).

42 Communications New Brunswick, 'New Brunswick government supports farming industry,' news release, 7 March 2002; and Communications New Brunswick, 'Department enacts new Agricultural Operations Practices Act,' news release, 10 January 2003.

43 CBC, 'Farmers across the province are celebrating a new Right to Farm Act,' nb.cbc.ca, 7 March 2002.

44 Morris, 'N.B. moves to protect farmers.'

45 The application, submitted in November 2005, remained unresolved as of May 2006. An agriculture ministry spokesperson refused to provide information about the application or the hearing, saying that the so-called public hearing would not be a 'publicized thing.' Lynn Moore, email to Elizabeth Brubaker, 23 May 2006, and telephone conversation with Elizabeth Brubaker, 29 May 2006. According to the ministry, staff 'effectively handled' the thirteen agricultural nuisance complaints received in 2003–4 and the ten received the following year. New Brunswick Ministry of Agriculture, Fisheries and Aquaculture, *Annual Report 2004–2005*, 23, http://www.gnb.ca/0168/10/2004-2005-e.pdf (accessed 21 November 2006).

46 Morris, 'N.B. moves to protect farmers.'

47 Daniel McHardie, 'N.B. ready to unveil Right to Farm Act,' *Moncton Times and Transcript*, 9 January 2003.

48 Janice Harvey, 'Regulations fall short of protecting environment and health,' *Saint John Telegraph-Journal*, 13 March 2002.

5. A Mushrooming Problem: Agricultural Nuisances in Ontario

1 Patricia Pyke, log entry, 19 April 1996. Cited in *Pyke v. Tri Gro Enterprises*, Ontario Superior Court of Justice, Court File No. 69190/95, 23 August 1999, 17.
2 Ibid.
3 The thirty-six mushroom producers in Langley Township, British Columbia, obtain their compost from two suppliers, both of which manufacture the material outside of the township. *T & T Mushroom Farm Ltd. v. Langley (Township)*, [1996] B.C.J. No. 30, New Westminster Registry No. S028796, paragraph 11.
4 D. Rinker, *Commercial Mushroom Production*, Publication 350, Queen's Printer for Ontario, 1993, 8. Cited in *Pyke v. Tri Gro Enterprises*, Ontario Superior Court of Justice, Court File No. 69190/95, August 23, 1999, 19.
5 The civil action was put on hold pending the resolution of issues before the Normal Farm Practices Protection Board.
6 Donald Dunn et al., *Ontario Right to Farm Advisory Committee Report*, submitted to the Honourable Jack Riddell, Minister of Agriculture and Food, 10 July 1986, 16–17.
7 Ibid., 21.
8 Ibid., 22.
9 Ibid., Appendix 4.
10 Ibid., 2, 5, 19.
11 Ontario Ministry of Agriculture, Food and Rural Affairs, *The Farm Practices Protection Act: Consultation Paper on the Role of the Farm Practices Protection Board*, February 1996, 2.
12 *Pyke v. Tri Gro Enterprises*, Ontario Superior Court of Justice, 29.
13 H.W. Fraser and F. Desir, *The Farming and Food Production Protection Act (FFPPA) and Nuisance Complaints*, Ontario Ministry of Agriculture, Food and Rural Affairs Factsheet, Order No. 03-113, published December 2003 and reviewed June 2004.
14 *Pyke v. Tri Gro Enterprises*, Ontario Superior Court of Justice, 24.
15 Ibid., 36.
16 *Pyke v. Tri Gro Enterprises*, Court of Appeal for Ontario, Docket C32764, 3 August 2001, 36.
17 *Gunby v. Mushroom Producers' Co-operative*, NFPPB File No. 1999-02. 31 C.E.L.R. (N.S.) 13 at 15.
18 Ibid., at 29.
19 *Gardner v. Greenwood Mushroom Farm*, NFPPB File No. 2000-01, 21 September 2000, 6, 9.

20 In 2004, GMF's neighbours asked the board to stay its proceedings pending the resolution of the private prosecution under the Provincial Offences Act. The board refused, insisting that it had concurrent jurisdiction. A challenge to that ruling was dismissed by the Divisional Court in September 2005. The board therefore proceeded with its investigation of GMF. A hearing in March 2006 led to negotiations between GMF and its neighbours – negotiations that were underway in May 2006, the time of this writing.

21 In October 2005, a justice of the peace dismissed GMF's motions to quash or stay the charges against it. A trial was scheduled for 2006.

22 *Pyke v. Tri Gro Enterprises*, Ontario Superior Court of Justice, 28.

23 One such challenge is described in note 20. A current court challenge concerns an NFPPB decision limiting a municipality's authority to set, in a zoning by-law, the minimum distance separating hog barns from other land uses. In March 2003, the board determined that a hog farmer's proposed expansion, if modified somewhat, would be considered normal and would not be subject to the minimum distance requirements in the municipality of Bluewater's zoning by-law. *Hill and Hill Farms Ltd. v. Municipality of Bluewater*, NFPPB File No. 2002-03. In January 2005, the Divisional Court ruled that the NFPPB has no jurisdiction to hear cases involving zoning by-laws. The Farming and Food Production Protection Act, it determined, allows the NFPPB to exempt normal farm practices only from municipal by-laws that prohibit or regulate nuisances. There is 'nothing in the Act to lead to the conclusion that the Board is qualified to deal with matters pertaining to land use planning.' *Hill and Hill Farms Ltd. v. Bluewater (Municipality of)* 74 O.R. (3d) 352. The Court of Appeal heard the case in March 2006, but had not issued its decision by May 2006, the time of this writing. Ontario Court of Appeal docket C43637.

24 Fraser and Desir, *Farming and Food Production*.

25 Michael Toombs, *Odour Control on Livestock and Poultry Farms*, Ontario Ministry of Agriculture and Food Factsheet, Order No. 03-111, December 2003, http://www.omafra.gov.on.ca/english/engineer/facts/03-111.htm (accessed 21 November 2006).

26 *Dietz v. Bigras*, NFPPB File No. 1997–04.

27 Fraser and Desir, *Farming and Food Production*.

28 *Knip v. Township of Biddulph*, NFPPB File No. 1998–02.

29 *Kelly v. Alderman*, NFPPB File No. 1997–02.

30 *Burns v. Perth South (Township) Chief Building Official*, 54 O.R. (3d) 266, [2001] O.J. No. 2117, paragraph 62.

31 *Thuss v. Shirley*, NFPPB File No. 1990-02; *McKenzie and Braeker v. Case Corporation and Stoltz Sales and Service*, NFPPB File No. 1996–01; and *Dietz v. Bigras*.

32 *Bader v. Dionis*, NFPPB File No. 1992–01; *Horbal v. Deschatelets*, NFPPB File No. 2000-03; and *Van Order v. Nolan*, NFPPB File No. 2003–02.

33 Wayne Caldwell, 'The Normal Farm Practices Protection Board,' *Final Report: Conflict Resolution in Rural Ontario*, June 2004, 9, http://www.waynecaldwell.ca/Projects/Conflict/right%20to%20farm%20practices.pdf (accessed 21 November 2006).

34 *Vivian v. Riar*, NFPPB File No. 1998-01.

35 *Parker v. Demmers*, NFPPB File No. 2001-08. The cost-sharing arrangement is mentioned not in the board decision but in the summary of the decision posted on the board's web site.

36 Untitled, NFPPB File No. 1989-01.

37 *Kelly v. Alderman.*

38 *Carson v. Werner*, NFPPB File No. 1997-03.

39 *Huff v. Prinzen*, NFPPB File No. 1990-01.

40 *Thuss v. Shirley.*

41 *Youcke v. Hermann*, NFPPB File No. 1993-01.

42 *Burns v. Perth South (Township) Chief Building Official*, paragraph 61.

43 *Kinrade v. Harrison*, NFPPB File No. 1997-01.

44 *Carson v. Werner.*

45 *Pyke v. Tri Gro Enterprises*, Ontario Superior Court of Justice, 29.

46 Agricultural Advisory Team, *Advice to the Government of Ontario*, October 2004, 8, http://www.omafra.gov.on.ca/english/aat/advice.pdf (accessed 21 November 2006).

47 Richard Mackie, 'Cattle producers to get $100-million in aid,' *Globe and Mail*, 28 September 2004.

6. Beyond the Right to Farm: Drainage and Planning Laws

1 Elbert van Donkersgoed, 'A problem defined does not equal a solution,' *Corner Post*, 18 March 2005.

2 R.W. Irwin, *Drainage Legislation*, Ontario Ministry of Agriculture and Food Factsheet, Order No. 89-166, revised 1997, http://www.omafra.gov.on.ca/english/engineer/facts/89-166.htm (accessed 21 November 2006).

3 Land Improvement Contractors of Ontario, *Ontario's Tile Drainage at 2000*, Factsheet No. 11, http://www.drainage.org/factsheets/fs11.htm (accessed 21 November 2006).

4 Ron Fleming, 'Human Health, Environmental Health, and Hog Manure,' in Proceedings of Swine Production and the Environment Seminar, *Living With Your Neighbours* (Shakespeare, ON, 1997), http://www.gocorn.net/v2006/Manure/articles/mag_manure2.htm (accessed 21 November 2006).

A study conducted by the Upper Thames Conservation Authority in Ontario found liquid manure applied to un-tilled fields at approved rates turned water in tile runs cloudy within five minutes and black within 15–20 minutes. It estimated that 2 per cent of the manure went down through the tiles. *West Perth (Township) Zoning By-law No. 100-1998 (Re)*, [2000] O.M.B.D. No. 707, File Nos. PL000064, R000010, 18 July 2000.

Brian Pett, reporting on a paper prepared for the Walkerton Inquiry by Michael Goss, notes that 'the most frequently reported type of manure spill is the movement of liquid manure, after it has been spread on land, to tile lines.' W. Brian Pett, *The Management of Manure and Non-Point Source Contamination of Water Quality in Ontario*, prepared for the Ontario Water Works Association and Ontario Municipal Water Association, August 2001, 7. In Ontario, between 1988 and 2001, 44 per cent of the MOE reported spills impacted field tile. Christine Brown, 'Recipe for Keeping Clean Water When Applying Liquid Manure?' Ontario Ministry of Agriculture, Food and Rural Affairs, June 2005, http://www.omafra.gov.on.ca/english/crops/field/news/croptalk/2005/ct_0605a3.htm (accessed 21 November 2006).

5 T.W. Van der Gulik et al., 'Managing Excess Water,' in *The Health of Our Water: Toward Sustainable Agriculture in Canada*, ed. D.R. Coote and L.J. Gregorich, 129.

6 Hugh Fraser and Sid Vander Veen, *Top 10 Common Law Drainage Problems between Rural Neighbours*, Ontario Ministry of Agriculture and Food Factsheet, Order No. 98-015, April 1998, http://www.omafra.gov.on.ca/english/engineer/facts/98-015.htm (accessed 21 November 2006). In 1974, an Alberta judge determined that farmers caused a nuisance when they built a berm along their property line. The berm blocked the flow of water draining from neighbouring lands, causing them to flood. The depressions through which the water had flowed (before the creation of the berm) formed a natural watercourse, which was entitled to protection under the common law. Having concluded that the law 'makes it quite clear that a lower proprietor on a natural watercourse has no right to block the water flowing therein, and, if he should do so, and thereby cause another person harm, he will be liable,' the judge awarded damages and issued an injunction requiring the defendants to remove the berm. *Kapicki, Ostashek, Tkachuk and Zabrick v. Andriuk and Andriuk (No. 2)*, [1975] 2 W.W.R. 264 at 276–7.

7 *Scott Rur. Mun. v. Edwards*, [1934] 3 D.L.R. 793 at 793–4.

8 Ibid., at 796.

9 *Smith v. Autoport Ltd.* (1973), 39 D.L.R. (3d) 248 at 258, 259, citing *Woolner v. Dyck*, [1950] 4 D.L.R. 745 at 748–9.

10 Ibid., at 260, citing *Hayden v. C.N.R. Co.* (1971), 16 D.L.R. (3d) 544 at 548–9.
11 In 2005, the agriculture ministry announced a new $6 million Agricultural
    Drainage Infrastructure Program under which the government would pro-
    vide grants to agricultural landowners for up to one third of the costs – two
    thirds of the costs in Northern Ontario – of drain construction and
    improvement projects. It also agreed to provide grants to municipalities to
    cover one half of the cost of employing a drainage superintendent. Ontario
    Ministry of Agriculture, Food and Rural Affairs, 'Ontario government sup-
    ports rural infrastructure,' news release, 22 September 2005. In the past,
    grants have been as high as 80 per cent in territories without municipal
    organization. Irwin, *Drainage Legislation*.
12 Cliff Evanitski, *The Drain Primer: Guide to Maintaining and Conserving Agri-
    cultural Drains and Fish Habitat*, Fisheries and Oceans Canada et al., created
    31 May 2002, updated 3 September 2003, http://www.dfo-mpo.gc.ca/
    canwaters-eauxcan/infocentre/guidelines-conseils/guides/drain-primer/
    drain1_e.asp (accessed 21 November 2006).
13 Van der Gulik et al., *Managing Excess Water*, 124.
14 Fraser and Vander Veen, *Top 10*.
15 Ted Cooper, emails to Elizabeth Brubaker, March–April, 2004.
16 Ted Cooper, email to Elizabeth Brubaker, 29 August 2003; and Ted Cooper,
    letter to the Ontario Ministry of the Environment, 31 March 2004.
17 Drainage is by no means the only farming practice exempted from
    Ontario's environmental protection legislation. The Environmental Protec-
    tion Act includes several exemptions for pollution from farms that follow
    normal practices. Such exemptions appear in sections 13 and 15, which
    require those who discharge excessive levels of contaminants to notify the
    environment ministry, and in part 10, which requires the restoration of the
    environment after a spill and the compensation of those who have been
    adversely affected. None of these provisions applies to 'animal wastes dis-
    posed of in accordance with normal farming practices.' A similar exemp-
    tion appears in section 14, which otherwise provides a broad prohibition
    against pollution: 'Despite any other provision of this Act or the regula-
    tions, no person shall discharge a contaminant or cause or permit the dis-
    charge of a contaminant into the natural environment that causes or is
    likely to cause an adverse effect.' Other exemptions for agriculture under
    the Environmental Protection Act fail to specify even that practices must be
    normal. Section 9, which requires a person to obtain a certificate of approval
    before constructing or altering any structure or equipment that may dis-
    charge a contaminant into the environment, does not apply to 'any plant,
    structure, equipment, apparatus, mechanism or thing used in agriculture.'

A more limited exemption can be found in section 6 of the Environmental Protection Act, which specifies that 'no person shall discharge into the natural environment any contaminant, and no person responsible for a source of contaminant shall permit the discharge into the natural environment of any contaminant from the source of the contaminant, in an amount, concentration or level in excess of that prescribed by the regulations.' The exemption for manure follows immediately. For more than a decade, the exemption applied to 'animal wastes disposed of in accordance with normal farming practices'; it now also requires the wastes to be disposed of in accordance with the Nutrient Management Act.

18 Angus Chuang, 'Taiwan Sugar moves hog farms offshore, eyes Canada,' Reuters, 6 January 2000.
19 Andrew Nikiforuk, 'The price of bringing home the bacon,' *Globe and Mail*, 12 June 2000.
20 Ralph Klein, letter to Maria Demers, 29 September 1998.
21 Legislative Assembly of Alberta, *Hansard*, 24 February 1999, 3:50, 3:40.
22 The Sustainable Management of the Livestock Industry in Alberta Committee, *Report and Recommendations*, 30 April 2001, 8, http://www1. agric.gov.ab.ca/$department/deptdocs.nsf/all/epw4866/$file/sustain. pdf?OpenElement (accessed 21 November 2006).
23 Legislative Assembly of Alberta, *Hansard*, 15 November 2001, 3:30.
24 Legislative Assembly of Alberta, *Hansard*, 20 November 2001, 8:30.
25 Legislative Assembly of Alberta, *Hansard*, 15 November 2001, 3:30.
26 Ibid., 4:30.
27 Ibid., 3:40–50.
28 Legislative Assembly of Alberta, *Hansard*, 27 November 2001, 5:20.
29 Legislative Assembly of Alberta, *Hansard*, 20 November 2001, 9:20.
30 Legislative Assembly of Alberta, *Hansard*, 15 November 2001, 4:30.
31 Society for Environmentally Responsible Livestock Operations (of Alberta), letter re: Bill 40, *Hog Watch Manitoba News*, April 2004.
32 Bryan Hill, letter to OMAFRA's Randy Jackiw, 18 April 1997.
33 Environmental Commissioner of Ontario, *Choosing our Legacy: 2003/2004 Annual Report – Supplement* (Toronto: ECO, 2004), 55.
34 Ontario Ministry of Agriculture, Food, and Rural Affairs, *Discussion Paper on Intensive Agricultural Operations in Rural Ontario*, 2000, http://www. omafra.gov.on.ca/english/agops/discussion.html (accessed 21 November 2006).
35 Konrad Yakabuski, 'High on the hog,' *R.O.B. Magazine*, September 2002, 70.
36 One challenge launched by the Ministry of Municipal Affairs concerned a zoning by-law for the municipality of West Perth, designed to restrict and

regulate intensive livestock operations. Government representatives contended that provisions in the by-law were 'arbitrary,' 'unnecessary,' and without basis in science or policy. The Ontario Municipal Board determined that it was within the jurisdiction of the municipality to limit the number of livestock at one site to 600 animal units and to require farmers to own 30 per cent of the land required for the spreading of livestock waste. *West Perth (Township) Zoning By-law No. 100-1998 (Re)*. The Divisional Court dismissed an appeal of the OMB ruling, confirming the municipality's authority to regulate and control extensive livestock production. In its decision, it referred at length to the Supreme Court of Canada's decision in *Spraytech v. Hudson*, which upheld a municipality's jurisdiction to pass a by-law limiting pesticide use. (See note 45, below.) Such a by-law, the Supreme Court found, did not conflict with provincial or federal regulation: 'A true and outright conflict can only be said to arise when one enactment compels what the other forbids.' *Ben Gardiner Farms Inc. v. West Perth (Township)*, [2001] O.J. No. 4394, Court File No. 1155, 7 November 2001. In March 2002, the Court of Appeal dismissed a subsequent appeal by the Ontario Cattlemen's Association.

37 *Hill and Hill Farms Ltd. v. Bluewater (Municipality of)* 74 O.R. (3d) 352; and Ontario Court of Appeal docket C43637. For more information, see page [000], note 23.

38 The Nutrient Management Act and its regulation concern the siting of buildings and storage facilities, the drainage of land, and the amount of land required for spreading manure. They restrict, to varying degrees, the application of manure near wells or surface water, on slopes, on saturated or frozen soils. They require the creation of five-year, site-specific manure management strategies or plans for approval by the Ministry of the Environment. Livestock farmers subject to the act must spell out how much manure they produce, what nutrients it contains, how it will be stored, how and where it will be spread, and how contingencies will be handled. For an overview of the Nutrient Management Act, see Anne Wordsworth, *Nutrient Management FAQs*, Resource Library for the Environment and the Law and Canadian Environmental Law Association, January 2004, http://www.cela.ca/faq/cltn_detail.shtml?x=1499 (accessed 21 November 2006).

39 The in-house counsel and senior policy advisor to the Environmental Commissioner of Ontario points out that the Nutrient Management Act is not the only law that claims to take precedence. The Planning Act likewise claims that, in the event of a conflict with other acts, the Planning Act is paramount. David McRobert and Lillian Hopkins, 'What Makes Nutrient Management So Controversial?' in *London Swine Conference: Building Blocks*

for the Future, April 2004, 95–6, http://www.londonswineconference.ca/
proceedings/2004/LSC2004_DMcRobert.pdf (accessed 21 November
2006).

The issue of supersedence has been before the courts, in the form of a
challenge to a Norfolk County by-law. The county, concerned that an
expanding hog operation would threaten local drinking water supplies,
wanted to enforce a by-law containing provisions more restrictive than
those found in the Nutrient Management Act. In a November 2004 ruling,
the Superior Court found that the NMA superseded the by-law; it
approved the hog farm's expansion. The county appealed. The Court of
Appeal heard the case in December 2005, but had not released its decision
by May 2006, the time of this writing.

Amendments to the Building Code Act have also given the Nutrient
Management Act an edge over municipal by-laws. Prior to 2003, building
officials were required to issue permits for proposed facilities if they would
not violate unspecified 'applicable law.' The government has since replaced
the general definition of 'applicable law' with a list of specific laws and regulations, including, for proposed livestock operations, the Nutrient Management Act. No longer will building officials have to verify compliance
with a variety of municipal by-laws before issuing permits.

In contrast, Ontario's Clean Water Act is expected to enable municipalities to better protect drinking water sources. Municipalities and Conservation Authorities will create, implement, and enforce source protection
plans. According to the Environmental Commissioner of Ontario, 'Source
protection will trump other concerns ... Instruments issued under a source
protection plan will have primacy over requirements of the Nutrient
Management Act.' *Neglecting Our Obligations: 2005/2006 Annual Report*
(Toronto: ECO, 2006), 24.

40 Legislative Assembly of Ontario, *Hansard*, 4 December 2001, 4197.
41 Ontario Ministry of Agriculture and Food, *The interaction between municipal
    by-laws and O. Reg 267/03 the general regulation made under the Nutrient Man-
    agement Act, 2002*, Factsheet, 30 January 2004.
42 Although Liberal MPP Lyn McLeod warned that 'one-size-fits-all policies
    are not going to do justice either to the environmental needs of nutrient
    management or to the realities that face farmers,' she also asserted a need
    for province-wide standards. Legislative Assembly of Ontario, *Hansard*,
    4 December 2001, 4204, and 11 December 2001, 4476.
43 Standing Committee on Justice and Social Policy, *Hansard*, 5 September
    2001, J-183. Also see ibid, J-204; and Legislative Assembly of Ontario, *Han-
    sard*, 4 December 2001, 4220.

44 Standing Committee on Justice and Social Policy, *Hansard*, 5 September 2001, J-201.

45 In 2001, the Supreme Court of Canada upheld the rights of municipal governments to restrict the use of pesticides within their boundaries. The case concerned a by-law passed a decade earlier by the Town of Hudson, Quebec. Two lawn care companies had challenged the by-law, which had restricted non-essential uses of pesticides in the community. (*114957 Canada Ltée (Spraytech, Société d'arrosage) v. Hudson (Town)*, 2001 SCC 40. File No. 26937.)

46 Jerry DeMarco, presentation to Standing Committee on Justice and Social Policy, *Hansard*, 5 September 2001, J-179.

47 Don Mills, presentation to Standing Committee on Justice and Social Policy, *Hansard*, 5 September 2001, J-200–1.

48 Christine Elwell et al., *Sixth Annual Report on Ontario's Environment* (Toronto: Canadian Institute for Environmental Law and Policy, January 2002) 33. Environmental organizations have also charged that the legislation was ineffective in controlling pollution. The Sierra Club of Canada accused the government of producing 'a "drip feed" of regulations that affect almost no farms, but that is "noisy" enough to make it look like they are doing something.' Maureen Reilly, 'The Ontario Nutrient Management Act – Sowing Confusion,' July 2003, email distributed by Ontario Environment Network. And the Environmental Commissioner of Ontario complained, 'since the majority of the rules will not apply to most farms in Ontario, until at least 2008, this regulation does little to reduce degradation of water quality due to nutrient management.' Environmental Commissioner of Ontario, *Choosing Our Legacy*, 77.

49 Clare Schlegel, presentation to Standing Committee on Justice and Social Policy, *Hansard*, 5 September 2001, J-181.

50 John Alderman and Franklin Kains, presentation to Standing Committee on Justice and Social Policy, *Hansard*, 10 September 2001, J-227–9.

51 Ontario Federation of Agriculture, 'New nutrient management legislation applauded,' *OFA Insider*, August/September 2001.

52 Jack Wilkinson, presentation to Standing Committee on Justice and Social Policy, *Hansard*, 5 September 2001, J-204.

53 Ontario Farm Environmental Coalition, letter to Ministers Steve Peters and Leona Dombrowsky, 1 December 2003.

54 Elbert van Donkersgoed, 'Rethinking the Regulatory Burden Swamping Agriculture,' *Corner Post*, 18 November 2004. In September 2005, in response to lobbying from the livestock industry, the provincial government amended its nutrient management regulation, relaxing requirements

that farmers submit manure management plans for approval and keep records of how they comply with those plans. Those changes, the Environmental Commissioner of Ontario charged, 'have weakened both accountability and the assurance that farmers are following the rules that protect human health' and 'may make key aspects of both the regulation and the Nutrient Management Act itself virtually unenforceable.' 'Backgrounder: 2005/2006 Annual Report,' 3 October 2006, 1.

55 Minister of Municipal Affairs John Gerretsen introducing Bill 26 in the Legislative Assembly of Ontario, *Hansard*, 15 December 2003. Other MPPs liberally used rhetoric about the importance of preserving local power. Liberal MPP Ernie Parsons commented, 'in many ways the best government is local government,' Legislative Assembly of Ontario, *Hansard*, 4 May 2004, 15:50, while his colleague Laurel Broten maintained, 'local knows local and Etobicoke knows Etobicoke.' Ministry of Municipal Affairs and Housing, transcript of town hall meeting / public information session on Planning Reform Act, Etobicoke, 17 June 2004, 2.

56 John Barber, 'Premier rights self after stumbling over Oak Ridges,' *Globe and Mail*, 17 February 2004.

57 Hill, letter to OMAFRA.

58 The OMB also lost its power to determine urban settlement area boundaries.

59 Several Progressive Conservative MPPs spoke out against the bill. Their opposition was hardly surprising, given that it had been the PC government that had introduced the requirement that municipal policies 'have regard to' provincial policy statements rather than comply with them. MPP Tim Hudak expressed 'great concern that local decision-making is taken away, in a number of substantive ways, under Bill 26.' Legislative Assembly of Ontario, *Hansard*, 4 May 2004, 17:40. His colleague John Baird objected to 'the centralization of power in what has become known as the politburo at the Ministry of Municipal Affairs.' Legislative Assembly of Ontario, *Hansard*, 12 May 2004, 18:00.

60 Many of the participants in the public consultation on proposed planning reforms were concerned primarily about urban sprawl and the loss of farm land and wild land. Many of these participants focused their attention on curbing the authority of the OMB, which they perceived not only as expensive, intimidating, and unaccountable but also as favouring developers.

61 Ontario Ministry of Municipal Affairs and Housing, 'Ontario government helps communities plan for the future,' news release, 21 February 2005.

62 Ontario Ministry of Municipal Affairs and Housing, 'New provincial policy statement,' Backgrounder, 21 February 2005.

63 Ontario Ministry of Municipal Affairs and Housing, *Provincial Policy State-ment*, 2005, 17, http://www.mah.gov.on.ca/userfiles/page_attachments/Library/1/789108_ppsenglish.pdf (accessed 22 November 2006).

64 Ed Tyrchniewicz et al., *Sustainable Livestock Development in Manitoba: Find-ing Common Ground*, report prepared for the Government of Manitoba by the Livestock Stewardship Panel, December 2000, 10, http://www.gov.mb.ca/agriculture/news/stewardship/livestock.pdf (accessed 13 November 2006).

65 MaryAnn Mihychuk, Legislative Assembly of Manitoba, *Hansard*, 14 April 2004, 15:30.

66 Leah Janzen, 'RMs to get more say on hog barns,' *Winnipeg Free Press*, 12 March 2004; and CBC, 'RMs to determine hog-barn locations in advance,' winnipeg.cbc.ca News, 12 March 2004.

67 Sean Maraj, 'Farmers' groups, RMs concerned by changes to Planning Act,' Portage la Prairie *Daily Graphic*, 16 April 2004.

68 Hogwatch Manitoba, 'Bill 40 – Planning Act Amendment – Discussion notes on some of the major changes,' 3 April 2004, 6, http://www.hogwatchmanitoba.org/bill40_planning_act_amendment_discussion_notes.htm (accessed 22 November 2006).

69 National Farmers Union, *Submission to the Government of Manitoba*, 15 April 2004, 10, 13.

70 Manitoba Pork Council, 'Government given bare passing grade for Plan-ning Act changes,' news release, 15 March 2004.

71 KAP passed two resolutions linking the Planning Act amendments to nor-mal farming practices. One read: 'That KAP lobby the provincial govern-ment to have Bill 40 include a clause stating that municipal by-laws cannot be applied to restrict a normal farm practice carried on as part of an agricul-tural operation.' The second read: 'That KAP lobby the provincial govern-ment to amend the Farm Practices Protection Act to allow an application, to the Farm Practices Protection Board, to be made by farmers who are directly affected by a municipal by-law that may have the effect of restrict-ing a normal farm practice.' Keystone Agricultural Producers, 'Resolutions from General Council,' *KAP Alert*, 16 April 2004.

72 'Hog barns on hold,' editorial, *Winnipeg Free Press*, 3 November 2004; and Province of Manitoba, 'Province moving ahead on land use planning,' news release, 2 November 2004.

73 Manitoba News Media Services, 'Modernized Planning Act would enhance public involvement,' news release, 25 April 2005; and Legislative Assembly of Manitoba, *Hansard*, 28 April 2005, 2076–7.

74 Legislative Assembly of Manitoba, ibid., 2084.

75 Telephone conversation with Elizabeth Brubaker, 16 June 2005.
76 Joe Dolecki, submission to Standing Committee on Legislative Affairs, *Hansard*, 6 June 2005, 275.
77 Carol Clegg, presentation to Standing Committee on Legislative Affairs, ibid., 231.
78 Glen Koroluk, presentation to Standing Committee on Legislative Affairs, *Hansard*, 7 June 2005, 298.
79 Clegg, presentation to Standing Committee, 232.
80 Peter Mah, presentation to Standing Committee on Legislative Affairs, *Hansard*, 7 June 2005, 300.
81 Larry Schweitzer, presentation to Standing Committee on Legislative Affairs, *Hansard*, 6 June 2005, 211–12.

7. Reversing the Trend: Decentralizing

1 The civil servant who handles all complaints and inquiries under the BC Farm Practices Board offered the following insight into the implications of using 'normal' as a standard for farmers' practices: 'What normal farm practice means, quoting from the legislation, is following "proper and accepted customs and standards as established and followed by similar farm businesses under similar circumstances." *So it's basically up to the industry to determine what the standard is,* because it's similar farms in similar circumstances. *If that farmer is doing what other farmers do, he or she is okay.* In addition to that, normal farm practice may also be determined by Lieutenant-Governor-in-Council standards.' Jim Collins, presentation to BC Select Standing Committee on Agriculture and Fisheries, *Hansard*, 25 May 1999, 16:35, emphasis added.

2 For more information on provincial right-to-farm laws, see Jonathan Kalmakoff, '"The Right to Farm": A Survey of Farm Practices Protection Legislation in Canada,' *Saskatchewan Law Review* 62 (1999): 225–68; Donald Good and Robert Paddick, *A Comprehensive Guide to Right to Farm in Canada*, background paper for workshop, Seeking Common Ground: Right to Farm and Private Land Use Issues (Ottawa: National Agriculture Environment Committee, 1996); George Penfold et al., 'Right-to-Farm in Canada,' *Plan Canada* 29.2 (1988): 47–55; and Charles Framingham, 'The Right to Farm – A Bandaid for a Heart Attack,' in *Law, Agriculture and the Farm Crisis*, ed. D.E. Buckingham and K. Norman (Saskatoon: Purich Publishing, 1992), 136–40.

3 According to one study of the inefficiency in Canada's farm economy, in 2001, as in previous years, farmers' earnings 'were exceeded by direct sub-

sidies. In addition to direct subsidies, Canada provides indirect subsidies to the agriculture sector. Over 10 years, the federal and provincial governments supplied approximately $3.53 in agricultural subsidies for every dollar earned by a Canadian farmer. No province operated a profitable farm economy over the past decade.' The authors note that the level of subsidies they report understates the direct and indirect financial assistance that Canadian farmers receive. Lawrence Solomon and Carrie Elliott, *Agricultural Subsidies in Canada 1992–2001* (Toronto: Urban Renaissance Institute, 2002).

4 Jack Wilkinson, President, Ontario Federation of Agriculture, 'Developing rules for intensive agriculture,' http://www.hurontel.on.ca/~heinzp/alert/ consult/ofalegiscoment.htm, accessed 2 November 2000. Elbert van Donkersgoed has campaigned vigorously for payments for the ecological services that farmers are asked to provide, including source water protection. 'Farmers know how to enhance their stewardship,' he wrote in one column. 'But without a payment system for the environmental goods and services that we already know can be part of the business of farming, little will happen.' In another column he insisted that pollution prevention 'is not a responsibility for a farmer alone. Those who eat cheap food as a result of the farming system should participate in the costs of providing a greater environmental assurance.' 'A strategy for sidestepping our present problems,' *Corner Post*, 11 October 2005; 'The service of environmental assurance, responding to the critics,' *Corner Post*, 20 March 2006. Agricultural lawyer Donald Good has likewise called for compensation for farmers whose practices are constrained in order to benefit the public: 'Where federal regulations and programs constrain the use of normal farm practices for the greater public good, a procedure [should] be developed to allow farmers to receive annual compensation payments for the interference.' Good and Paddick, *A Comphrehensive Guide*, 47.

5 Ronald Coase, winner of the 1991 Nobel Prize in economics and author of 'The Problem of Social Cost,' the most cited law-review article in history, famously pointed out that, in some cases, a farmer's neighbour may pay him to cease operating in a particular manner or place. Under the Coase Theorem, regardless of who initially holds the rights to use a disputed resource, the right will be acquired by the party that values it most, assuming that transaction costs are not prohibitive. 'If market transactions were costless,' Coase wrote, 'all that matters (questions of equity apart) is that the rights of the various parties should be well-defined.' Ronald Coase, 'The Problem of Social Cost,' *Journal of Law and Economics* 3 (October 1960): 19. Transaction costs, however, are almost never zero. Indeed, they can be

extensive, especially if many parties are involved. Gathering information, identifying, communicating with, and organizing affected parties, negotiating agreements, monitoring results, documenting infractions, and enforcing corrective action can increase transaction costs and prevent market-style bargaining. Nor are questions of equity 'apart.' Although economists may be satisfied with the promise of an efficient outcome, individual participants will not be so sanguine. Who holds the contested rights – and who has to pay whom to purchase them – will affect the distribution of wealth between the parties. It may also offend the parties' sense of moral and legal justice to have to purchase rights that they feel are rightfully theirs. Murray Rothbard, 'Law, Property Rights, and Air Pollution,' *Cato Journal* 2.1 (spring 1982): 57–60; and Gary North, 'Undermining Property Rights: Coase and Becker,' *Journal of Libertarian Studies*, 16.4 (fall 2002): 75–90.

6 *Pyke v. Tri Gro Enterprises*, Court of Appeal for Ontario, Docket C32764, 3 August 2001, 42–3.

7 *Bormann v. Board of Supervisors in and for Kossuth County*, 584 N.W.2d 309, 29 Envtl. L. Rep. 20,235 (Iowa, Sep 23, 1998) (No. 192, 96–2276). The U.S. Supreme Court refused, without comment, to review this ruling.

8 *Imperial Oil Ltd. v. Quebec (Minister of the Environment)*, 2003 SCC 58. Following its meeting in Rio de Janeiro in 1992, the UN Conference on Environment and Development proclaimed that 'the polluter should, in principle, bear the cost of pollution' and 'national authorities should endeavour to promote the internalization of environmental costs.' *Rio Declaration*, Principle 16, http://www.unep.org/Documents.multilingual/Default.asp?DocumentID=78&ArticleID=1163 (accessed 22 November 2006). The polluter-pay principle is by now, according to Environment Commissioner Margot Wallstrom, 'a cornerstone of EU environmental policy.' David Hopkins, 'Polluter pays agreement on liability reached,' *Edie Weekly Summaries*, 27 February 2004.

9 Martin S. Beaulieu, 'Manure Management in Canada,' in Statistics Canada, *Farm Environmental Management in Canada*, 2004, vol. 1, no. 2 (catalogue no. 21-021-MIE – No. 002).

10 Wendy Powers, *Practices to Reduce Odor from Livestock Operations*, Iowa State University Extension, PM1970a, July 2004, 4; and Janelle Hope Robbins, *Understanding Alternative Technologies for Animal Waste Treatment: A Citizen's Guide to Manure Treatment Technologies* (Terrytown, NY: Waterkeeper Alliance, 2005), 20.

11 Penfold et al., 'Right-to-Farm in Canada,' 53–4.

12 Kalmakoff, 'The Right to Farm,' 255.

13 Natural Resources Conservation Board, *Q4 / 2003–04 Quarterly Report:*

Applications, Compliance, Reviews, 31 March 2004. The number of complaints declined the following year, and again the year after that. In 2005–6, the board received 878 complaints, 507 of which concerned odours or other nuisances created by 121 confined feeding operations. Garbine Lazar, NRCB, email to Elizabeth Brubaker, 31 May 2006.

14 Ontario Ministry of Agriculture, Food and Rural Affairs, *Farming and Food Production Protection Act: Information Summary*, 1 November 1999; Michael Toombs, *Odour Control on Livestock and Poultry Farms*, Ontario Ministry of Agriculture and Food Factsheet, Order No. 03-111, December 2003; and Normal Farm Practices Protection Board, *Annual Report of the Normal Farm Practices Protection Board – April 1, 2003 to March 31, 2004*, http://www. omafra.gov.on.ca/english/engineer/nfppb/annual_report2003.htm (accessed 22 November 2006). Fewer complainants seem now to be turning to the Normal Farm Practices Protection Board to resolve their concerns, perhaps because they have lost confidence in the board, or, alternatively, perhaps because they are satisfied by the conflict resolution process formalized by the board in 2005.

15 Good and Paddick, *A Comprehensive Guide*, iii. In 1986, the Ontario Right to Farm Advisory Committee found that neighbouring farmers were responsible for the majority of nuisance complaints. Donald Dunn et al., *Ontario Right to Farm Advisory Committee Report*, July 1986, 16. Approximately half of the complaints received by the agricultural ministry in 1994–5 came from farmers. Bernard Tobin, 'Right to farm rules under review,' *Farm and Country*, 14 May 1996.

16 Gary J. McTavish and David Lee, *Community Relations in the Rural Area*, Ontario Ministry of Agriculture and Food Factsheet, Order No. 98-065, November 1998.

17 Some jurisdictions subject large farms to more stringent environmental regulation than small farms. Perhaps for this reason, or perhaps because large farms have greater access to capital, technology, and expertise, some large farms use better practices than do their smaller counterparts. For example, 65 per cent of large hog farms have a manure storage capacity of more than 250 days, while just 41 per cent of small hog farms are able to store their manure for that long. Their increased storage capacity enables large farms to avoid winter spreading of manure. Just 2 per cent of the manure produced on large farms is spread in the winter, while 5 per cent of the manure produced on small hog farms is spread in the winter. Furthermore, large farms are more likely to inject or rapidly incorporate manure into the soil. (Lucie Bourque and Robert Koroluk, 'Manure Storage in Canada,' Statistics Canada, *Farm Environmental Management in Canada*, 2003, vol. 1, no. 1 (cata-

logue no. 21-021-MIE2003001), September 2003; and Beaulieu, 'Manure Management.'

18 For a more extensive discussion of this issue, see chapter 2.

19 It is impossible to know how often farmers have, through bargaining, resolved disputes with their neighbours, since informal arrangements – unlike court decisions – are not part of the public record and since success-ful agreements would have kept disputes from reaching the courts. Occa-sionally, court decisions refer to bargaining. For example, in a 1928 Ontario decision regarding noise from a dairy, the judge mentioned that the defen-dant had offered to purchase from one of the plaintiffs land that might have enabled it to avoid creating a nuisance, but that the two parties could not agree on a price. *Duchman v. Oakland Dairy Co. Ltd*, Ontario Law Reports, vol. 63, 116. For a discussion of the markets for environmental rights that have developed under the common law, see Bruce Yandle, *Common Sense and Common Law for the Environment: Creating Wealth in Hummingbird Econo-mies* (Lanham, MD: Rowman and Littlefield, 1997), 87, 98, 101, 106, 117.

20 Bruce Yandle argued that 'few environmental problems have dimensions that coincide with political units ... Most problems to be solved surely should be addressed at the level of small ecological units ... River basins or watersheds form natural units for managing water quality.' *Common Sense and Common Law*, 25.

21 William Weida, *Considering the Rationales for Factory Farming*, Global Resource Action Center for the Environment, March 2004, 38, http://www.factoryfarm.org/docs/Foundations_of_Sand.pdf (accessed 22 November 2006); William Weida, *A Summary of the Regional Economic Effects of CAFOs*, Global Resource Action Center for the Environment, July 2001, http://www.factoryfarm.org/docs/RegionalEcon72101. pdf (accessed 22 November 2006); and William Weida, *The Evidence for Property Devaluation Due to the Proximity to CAFOs*, Global Resource Action Center for the Environment, January 2002, http://www.factoryfarm.org/docs/Weida_Prop_Devaluation.pdf (accessed 22 November 2006). Livestock farms do not always lower nearby property values. A Manitoba study found that prices of houses within one or two miles of hog farms did not differ significantly from those within four or five miles; furthermore, land values generally increased in close proximity to hog farms. Royal LePage Stevenson Advisors, *Impact Analysis of Intensive Livestock Operations on Man-itoba Rural Residential Property Values: Five Case Study Locations*, prepared for Manitoba Pork Council, spring 2004.

22 One policy analyst called local governments 'the most accountable author-ity on the face of the Earth,' arguing that they face elections at fixed dates,

function without parties and backbenches, are accessible to voters, and are accountable not only to the electorate but also to their province. Casey Vander Ploeg, 'New deal for cities is no "power grab,"' *Financial Post*, 9 July 2005.

23 Elinor Ostrom and Edella Schlager have stressed the benefits of locally devised rules, explaining that external authorities 'lack the commitment to ensure their viability and longevity.' Margaret McKean has likewise noted the problems that arise when decision makers 'can insulate themselves from the undesirable social consequences of the actions.' Cited by Yandle, *Common Sense and Common Law*, 9, 25.

24 For more on the evolutionary nature of common-law environmental controls, see Yandle, *Common Sense and Common Law*, xv, 88, 111–12, 160, 170–1. The author pointed out that 'legal rule mutation is slowed by statute; the risk of survival of a flawed system rises' (8).

25 Christian Farmers Federation of Ontario, 'Guidance for large and/or intensive livestock enterprises,' policy statement, November 2002.

26 John Cotter, 'Environmental group calls on Ottawa to regulate Canada's factory farms,' Canadian Press, 3 October 2002.

27 Robert Benzie, 'New Ontario farm rules to protect water supply,' *National Post*, 21 August 2002.

28 Naomi Klein, 'What happened to the new left?' *Globe and Mail*, 30 January 2003.

29 Kevin Dougherty, 'Let regions set limits on pork production: Liberals,' *Montreal Gazette*, 1 May 2002.

30 R.A. Fowler, letter to Douglas Galt, MPP, 17 February 1999.

31 Lisa Ambus, 'Developing Sustainability through Local Control,' *Polis* (winter 2002/3): 10. For more on the importance of decentralized decision making, see Oliver Brandes et al., *At a Watershed* (Victoria: Polis Project, 2005), 17–19.

32 H2infO List Serv, 'Source Water Protection,' forwarded from Canadian Environmental Law Association, 12 June 2003.

33 Canadian Environmental Law Association, 'Victory for pesticide reduction and local democracy,' press release, 28 June 2001. CELA's confidence in municipalities is not unlimited. It has warned that municipalities may not have the political will to monitor industrial farms. Martin Mittelstaedt, 'Farm law aimed at water protection,' *Globe and Mail*, 24 January 2001.

34 Great Lakes United et al., 'Great Lakes groups endorse state and local leadership of Great Lakes recovery plan, call for reform of U.S. Army Corps of Engineers,' statement presented at Moving Toward a Sustainable Great Lakes conference, Sault Ste Marie, Michigan, 25 June 2003. In the context of

manure management, the Sierra Club has called the disempowering of municipalities inappropriate. Vik Kirsch, 'Farmers, environmentalists divided over effectiveness of new manure law,' *Guelph Mercury*, 28 June 2002.

35 Robert Remington and Charlie Gillis, 'Klein, Kennedy clash over factory farms,' *National Post*, 19 March 2002.

36 Kelly Cryderman, 'Kennedy, Klein clash over agency,' *Edmonton Journal*, 19 March 2002.

37 Linda Slobodian, 'Kennedy invites Klein to see hog farm damage,' *Calgary Herald*, 18 March 2002.

38 Pope Pius XI, *The Fortieth Year*, Encyclical on Reconstruction of the Social Order, 1931, paragraph 79, http://www.vatican.va/holy_father/pius_xi/encyclicals/documents/hf_p-xi_enc_19310515_quadragesimo-anno_en.html (accessed 22 November 2006).

39 U.S. Catholic Bishops, *Economic Justice for All*, pastoral letter on Catholic Social Teaching and the U.S. Economy, 1986, http://www.osjspm.org/economic_justice_for_all.aspx (accessed 23 November 2006).

40 Pope John Paul II, *Centesimus Annus* (*The Hundredth Year*), encyclical, 1991, http://www.vatican.va/holy_father/john_paul_ii/encyclicals/documents/hf_jp-ii_enc_01051991_centesimus-annus_en.html (accessed 23 November 2006).

41 Robert Vischer, 'Subsidiarity as Subversion: Local Power, Legal Norms, and the Liberal State,' 2, undated essay based on a chapter prepared for *Self-Evident Truths: Catholic Perspectives on American Law*, ed. Michael Scaperlanda and Teresa Collett, forthcoming.

42 David Bosnich, 'The principle of subsidiarity,' *Religion and Liberty*, July–August 1996.

43 *Treaty Establishing the European Community*, consolidated text, Official Journal C 325, 24 December 2002, article 5, http://europa.eu.int/eur-lex/en/treaties/dat/C_2002325EN.003301.html (accessed 23 November 2006).

44 Europa Glossary, http://europa.eu.int/scadplus/leg/en/cig/g4000s.htm (accessed 19 November 2004).

45 *114957 Canada Ltée (Spraytech, Société d'arrosage) v. Hudson (Town)*, 2001 SCC 40. File No. 26937.

46 Good and Paddick, *A Comprehensive Guide*, 35, 37.

47 John Swaigen, 'The "Right-to-Farm" Movement and Environmental Protection,' *Canadian Environmental Law Reports* 4 C.E.L.R. (N.S.), 126; Kalmakoff, 'The Right to Farm,' 233–4; Good and Paddick, *A Comprehensive Guide*, 8; and Penfold et al., 'Right-to-Farm in Canada,' 48, 51–2.

48 Eldon McAfee, 'Environmental Updates for Iowa Pork Producers,' *Iowa Pork Producer* 42.7–8 (October 2005): 9.
49 *Blass et al. v. Iowa Select Farms L.P.*, No. LACV018147 (Sac County, Iowa, Dist. Ct.).
50 Eldon McAfee, 'Nuisance Lawsuits – Living with the Risk,' *Iowa Pork Producer* 42.6–7 (July 2005): 11–12; and 'A Look at Nuisance Cases – from a Livestock Perspective,' *Iowa Pork Producer* 41.6–7 (July 2004): 15.
51 Sheldon Alberts and Bill Curry, 'New mad cow case threatens border reopening,' *National Post*, 12 January 2005; and editorial, 'A sane response to mad cow,' *National Post*, 14 January 2005. A class-action lawsuit filed against the federal government and a feed company, alleging negligence in failing to prevent the BSE crisis, claimed more than $7 billion in damages, an amount reflective of the income lost after the border closed. Sutton Eaves, 'Farmers to sue Ottawa over mad cow losses,' *National Post*, 12 April 2005. A $7–billion estimate also appeared in Cyril Doll, 'Feeding misconceptions,' *Western Standard*, 13 February 2006.
52 Mark Hume, 'Bird-flu fear sparks huge kill,' *Globe and Mail*, 6 April 2004; and Tim Naumetz and Sean Gordon, 'Ottawa orders vast B.C. bird kill,' *National Post*, 6 April 2004.
53 Editorial, 'Preparing for the next flu pandemic,' *National Post*, 23 February 2005.
54 Among the most important actions to be taken by upper-level governments are the elimination of the myriad subsidies that have exacerbated the environmental problems that now demand regulation.
55 A policy fellow with the Frontier Centre noted that under current one-size-fits-all regulations, a farmer with a manure lagoon on heavy clay soil far from any aquifer or stream, with more than enough land to inject manure into, faces the same requirements for approval and periodic inspection as does one who farms on top of an aquifer and beside a stream, with little land on which to spread his manure. Such regulations do not merely create unnecessary costs; they also create 'as much disincentive to put these things in the right place as in the wrong place.' Rolf Penner, email to Elizabeth Brubaker, 26 August 2005.
   Economists examining the use of economic instruments in agricultural regulation confronted farms' spatial, temporal, and technological heterogeneity and concluded that 'cost-effectiveness requires policy instruments to be targeted to individual farms.' They advocated 'a system of locally specified management incentives.' Alfons Weersink et al., 'Economic Instruments and Environmental Policy in Agriculture,' *Canadian Public Policy* 24.3 (1998): 313, 324. A water resources economist likewise noted, 'because of

the crucial importance in physical conditions (such as local soil types, groundwater-surface water interactions and weather conditions), the analysis of nonpoint source pollution and the design of policies aimed at controlling it in a least-cost fashion are likely to be quite case-specific.' Steven Renzetti, *Canadian Agricultural Water Use and Management*, Brock University Department of Economics Working Paper, March 2005, 12.

Index

**The University of Toronto Centre for Public Management
Monograph Series**